HYDROGEOLO
IN PRACTICE

A Guide to Characterizing Ground-Water Systems

William J. Stone
Now at Los Alamos National Laboratory

PRENTICE HALL
Upper Saddle River, New Jersey 07458

Stone, William J. (William Jay)
Hydrogeology in practice : a guide to characterizing ground-water systems / William J. Stone.
p. cm.
Includes bibliographical references and index.
ISBN 0-13-899154-5 (pbk.)
1. Hydrogeology. I. Title.
GB1003.2.S76 1999
551.49—dc21 98-41441
CIP

Executive Editor: ***Robert A. McConnin***
Production Editor: ***Joanne Hakim Jimenez***
Assistant Managing Editor: ***Lisa Kinne***
Manufacturing Manager: ***Trudy Pisciotti***
Art Director: ***Jayne Conte***
Cover Designer: ***Bruce Kenselaar***
Cover: Water quality (specific conductance) and flow direction of ground water in the Morrison Formation, San Juan Basin, New Mexico. Modified from W. J. Stone 1989. Hydrology of the San Juan Basin, in Finch, W. J., Huffman, A. C., Jr., and J. E. Fassett, eds. Coal, uranium, oil, and gas in Mesozoic rocks of the San Juan Basin—anatomy of a giant energy-rich basin: Field Trip Guidebook T120, p. 39–41. American Geophysical Union.
Sources of Part Opening Art: *Parts I, II, and III: Meinzer, O. E. 1923. The occurrence of ground water in the United States—with a discussion of principles. U.S. Geological Survey, Water-Supply Paper 489. 321 p. Part IV: Bowman, I. 1911. Well drilling methods. U.S. Geological Survey, Water-Supply Paper 257. 139 p.*
All photographs are by the author, except as otherwise noted.

Printed in the United States of America

10 9 8 7 6 5 4 3 2 1

ISBN 0-13-899154-5

Prentice-Hall International (UK) Limited, *London*
Prentice-Hall of Australia Pty. Limited, *Sydney*
Prentice-Hall Canada Inc., *Toronto*
Prentice-Hall Hispanoamericana, S.A., *Mexico*
Prentice-Hall of India Private Limited, *New Delhi*
Prentice-Hall of Japan, Inc., *Tokyo*
Simon & Schuster Asia Pte. Ltd., *Singapore*
Editora Prentice-Hall do Brasil, Ltda., *Rio de Janeiro*

Contents

Preface

At some point everyone pauses and reflects on his chosen field and career. What have we learned? Is any of it worth passing on to others? Also, we may be overcome by a feeling that students graduating today, and even some of the professionals already in the work force, have not been trained the same as we were, to the detriment of the field. There is even a temptation to try to synthesize all those books and papers on our office shelves (or in boxes in the garage).

This book grew partly out of these impulses, but mainly from the observation that many reports prepared by hydroscientists today: (1) are devoid of some of the most basic geologic and hydrologic information; (2) seem to be based on studies conducted without the benefit of a sound conceptual hydrogeologic model; and (3) are poorly written. This book is by no means intended to replace the numerous basic texts on hydrogeology. Rather, it is offered as: (1) a supplement to those texts used in applied hydrogeology courses; (2) a core text to be complemented by additional case histories for a course on characterizing hydrologic systems; and (3) a ready reference for professionals to consult when conducting hydrogeologic studies.

This book assumes the reader has some background in both geology and hydrology. However, with the general definitions of terms in the text and the glossary at the back, most students or professionals should be able to follow the discussion. Others may need to refer to standard geology or hydrology texts for more detailed explanations of basic concepts. Suitable geology and hydrology sources are included in the references.

Many people have contributed both directly and indirectly to the completion of this book. Discussions over the years with Elmer Baltz (when with the U. S. Geological Survey), John Hawley (when a colleague at the New Mexico Bureau of Mines and Mineral Resources), and Kelly Summers (both when a consultant and when with the City of Albuquerque) helped formulate my standards for hydrogeologic studies. Some material comes from courses I taught at the New Mexico Institute of Mining and Technology and the University of New Mexico. Questioning by students provided important feedback. I was encouraged throughout the project by various colleagues at the New Mexico Bureau of Mines and Mineral Resources and New Mexico Environment Department, some of whom looked at selected draft chapters. The book benefited most, however, from the comprehensive review by Michael E. Campana (University of New Mexico). While his efforts are most appreciated, the responsibility for any errors rests solely with me. Finally, I would like to thank my friend (and wife) Rosanna and my son Ian for their understanding and tolerance during my preoccupation with the preparation of the book.

In spite of a conscious effort to make the book comprehensive, it is no doubt incomplete. It was my intention to cover the basic elements of hydrogeologic studies. I hope that the book will serve at least as a starting point for considering this fascinating topic.

William J. Stone

This book is dedicated to my parents.

CHAPTER 1

Introduction

A spill was suspected and reported. A site investigation was made and the area was found to be contaminated. The consulting firm presented its findings in a timely manner. However, the report included no geologic map, cross sections or well-records table. The water-level map did not distinguish between the shallow unconfined and the deeper confined systems. Maps that were intended to show the extent and level of contamination mixed the concepts of contours and zones, and were not labeled appropriately for either. Symbols and abbreviations used were not explained in the legend. No attempt was made to present a conceptual hydrogeologic model of the area, let alone tie the fate of the contamination to it. In short, the report, and apparently the study upon which it was based, were unsatisfactory.

Unfortunately, this is all too common. Although this scenario is hypothetical, it was constructed from weaknesses found in actual reports. The example represents a remediation case, but the same concerns apply to water-resource and other types of hydrogeologic studies and reports.

What can be done? Since these reports become public documents, especially in the case of regulatory or legal action, they may be seen by a wide range of people. Surely it must be realized that the quality of such reports reflects on the competence of those under whose letterhead the reports are submitted. Thus, we can only conclude that these important items are omitted from reports because the technical staff members who prepare them and those that review their work before it is released (1) don't know what such reports should include, (2) don't know where to get the information required, or (3) don't know what to do with the information once they get it. Perhaps it is a combination of all three. It doesn't have to be that way. Certainly, as we approach the twenty-first century, we have the means to do better. This book was written in an attempt to help those involved in the water-resource and environmental industries improve their studies and reports.

HYDROGEOLOGIC STUDIES

The number of hydrologic problems recognized in the world increases every day. The problem may be one of too little water: inadequate quantity or quality of surface water or ground water (Figure 1-1). Conversely, the problem may be one of too much water, as in controlling surface runoff or ground-water inflow at a mine or construction site (Figure 1-2). Or the problem may be where to dispose of hazardous waste without impairing the local water supply. In places where these concerns were not carefully considered, or where agricultural chemicals or the contents of leaking pipelines, storage tanks, waste ponds, etc., have migrated to the water table or closest stream, the problem may be one of cleaning up the soil and water as much as possible (Figure 1-3).

FIGURE 1-1 Water shortage is often indicated by abandoned wells as in this old, dry, hand-dug well on the Navajo Indian Reservation, northwest New Mexico, USA.

FIGURE 1-2 Construction projects often require extensive dewatering, as in the rebuilding of a major diversion structure for the Middle Rio Grande Conservancy District, San Acacio, New Mexico, USA.

FIGURE 1-3 A reasonable understanding of the hydrogeologic setting is essential for remediating spill sites. Here, a "forest" of monitoring wells has been installed to provide information at a Superfund site in Albuquerque, New Mexico, USA.

All too often, environmental scientists or engineers rush to define a plume without first adequately researching and studying the geology and hydrology of the area. Addressing such problems requires a clear understanding of the geologic framework, the hydrologic system, and their relationship in the study area. Such understanding is generally provided by a hydrogeologic study. There are two sides to hydrogeology: theoretical and practical (Figure 1-4). Both are important.

Hydrogeology In Theory

The theoretical side of hydrogeology is the foundation upon which modern studies rest. It includes all that we know of water in nature from years of observation, experimentation, and deduction. More specifically, theoretical hydrogeology comprises the fundamental geologic principles as well as the basic laws and equations governing surface-water, soil-water, and ground-water behavior taught in the classroom. Thus, it includes such basic principles as the Manning equation (relating stream velocity and channel roughness), Henry's law (relating concentration of a gas in water to its pressure), Darcy's law (relating soil- and ground-water flow to conductivity, gradient, and area), and many others.

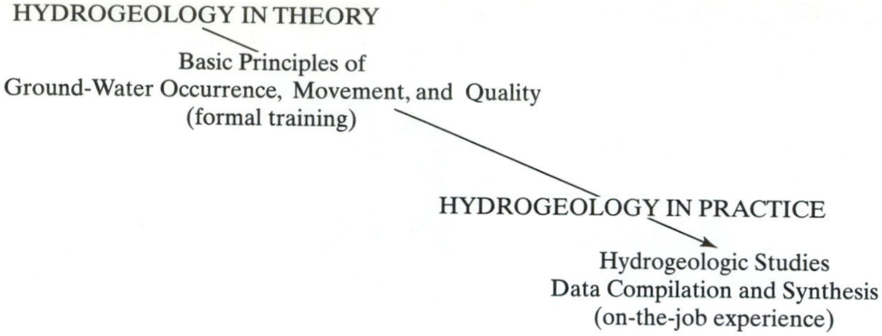

HYDROGEOLOGY IN THEORY

Basic Principles of
Ground-Water Occurrence, Movement, and Quality
(formal training)

HYDROGEOLOGY IN PRACTICE

Hydrogeologic Studies
Data Compilation and Synthesis
(on-the-job experience)

FIGURE 1-4 Relationship between hydrogeology in theory and in practice.

Hydrogeology In Practice

Although understanding theoretical concepts is essential to understanding hydrologic systems, it is the practical side of hydrogeology that we consider here. Practical hydrogeology involves the application of these theoretical principles to everyday problems. In times past, hydrogeologists rarely learned this skill in the classroom. Rather, they learned it through association with experienced professionals on the job. While there is no substitute for such training, the purpose of this book is to supplement it by providing some guidance on what hydrogeology consists of in practice.

A hydrogeologic study should be approached in a series of logical steps: (1) compiling the basic geologic and hydrologic information, (2) characterizing the geologic setting and hydrologic system, and (3) synthesizing this information into a coherent picture. Compiling information includes making a literature search for major references on the area under study, sifting through these for pertinent hydrogeologic data, and creating appropriate tables and illustrations incorporating these data. Characterizing consists of thoroughly describing the important hydrogeologic features and conditions of the area, based on the data compiled. Synthesizing involves formulating a sound conceptual hydrogeologic model, possibly testing it with a numerical model, and ultimately presenting it in a professional report.

SCOPE OF THIS BOOK

The scope of this book is narrow; it is not an introduction to either geology or hydrology. Neither is it intended to cover everything a water scientist should know. Rather, it is offered as a supplement to formal training and standard texts. It is offered as a quick reference on, or a reminder of, basic components of a sound hydrogeologic study.

The book was prepared with three audiences in mind: (1) seasoned professionals who, owing to pressures of budget and time constraints, have gotten into the habit of omitting some of the basic elements in their studies; (2) transplanted professionals—that is, those not trained in hydrogeology, but now working in water-resource or environmental fields—who need a practical guide to conducting hydrogeologic studies; (3) students who want additional information on putting theoretical hydrogeology into practice. Put another way, the book is aimed at those long out of school who knew what to do but forgot, others long out of school who never knew what to do but are seeking guidance, and those still in school and learning what to do.

ORGANIZATION AND CONTENTS

This book is divided into four parts. Part I reviews how geologic characterizations should be made. Part II covers the characterization of hydrologic systems. In part III the two concepts are combined, first in terms of a reasonable conceptual model and then as a sound hydrogeologic report. Part IV offers applications of these practices to the main types of studies confronting hydrogeologists today, as well as some final suggestions for hydroscience professionals. Although each chapter is designed to build upon the previous one, it should be possible to use any of them more or less independently, as well.

The book contains several means of locating information. The works listed at the end of the chapters are the sources of principles or case histories presented. These external sources may be consulted for more details or still further references on the given subject. There are also aids for quickly locating subjects within the book. The table of contents refers the reader to general subjects, whereas the index identifies where more specific topics are discussed.

SOME TERMS

Like any technical field, hydrogeology has its own unique terms. The most common terms are defined in the glossary at the back of the book. However, the meanings of a few major words and concepts are given at the outset for clarification.

There has been some debate as to whether *ground water* should be spelled as one or two words. In North America, the two-word form seems logical in view of the parallel term *surface water,* especially as the two may appear together in reports. In Europe, however, the single word has been favored, perhaps from the German *grundwasser.* Though a seemingly trivial matter, many hydrogeologists became involved in the pursuit of a solution. An informal poll of the state geological surveys by the Wisconsin Geological and Natural History Survey in 1986 revealed that all but 10 of the 46 states responding preferred the two-word form (hyphenated only when used as a unit modifier). This form is standard for reports of the U.S. Geological Survey (USGS), the oldest, largest, and most prolific publisher of hydrologic work, both nationally and internationally. Nonetheless, the one-word form was used in Webster's *Third International Dictionary* in 1974. When the USGS questioned this usage, the publishers conceded that citational evidence confirms a preference for the two-word form and promised to restore it in future editions. The two-word form is also used by the National Ground Water Association, whose journal *Ground Water* is one of the leading technical journals in the field. In spite of this body of evidence, the debate continues. The two-word form is used throughout this book, following these precedents.

There is also considerable confusion about the meaning of several terms applied to the various fields of hydroscience. This is due largely to a general lack of standard usage and to erroneous definitions, even by those who should know better. The following discussion is offered to clarify the meaning of major terms.

Hydrology is the broadest term used and covers the entire spectrum of scientific study of the earth's water (Meinzer 1923). As noted by Ward (1967), regarding hydrology as a branch of civil engineering is far too narrow a view. An understanding of water is critical to and an integral part of many other fields, such as agriculture, forestry, geology, soil science, etc. Some workers, especially those new to the field, consider that the term refers strictly to surface water. However, that realm is described by *hydrography,* the description of aqueous portions of the earth's surface (Howell 1962) or *potamology,* the study of streams (Gray 1973). *Surface-water hydrology* is more commonly used for the scientific study of lakes and streams. *Ground-water hydrology* is the scientific study of subterranean waters.

Most differences of opinion have centered on the terms *hydrogeology* and *geohydrology*. An early American Geological Institute (AGI) glossary (Howell 1962) amazingly showed them both to be equivalent to *hydrology;* in fact, this was given as the preferred meaning for *hydrogeology*. Todd (1980) noted that both were identical to *ground-water hydrology,* with *hydrogeology* differing only in its greater emphasis on geology. The USGS seems to favor this approach, using the term *geohydrology* almost exclusively for its ground-water studies.

Today, unfortunately, many people take the two terms to be interchangeable. May (1976) objected to this and, in a letter to the editor of *Geotimes,* attempted to clarify the differences between the terms. He argued that since English generally places the adjective before the noun it is modifying, *hydrogeology* is a branch of geology (the geology of water) and *geohydrology* is a branch of hydrology (the role of water in geologic processes). More specifically, May offered what he felt to be more correct and appropriate definitions for the two terms:

> Hydrogeology—the science that applies geologic methods to the understanding of hydrologic phenomena (following the lead of "hydrodynamics," etc.).
>
> Geohydrology—the science that applies hydrologic methods to the understanding of geologic phenomena (following the lead of "geochemistry," "geophysics," etc).

He further noted that both terms apply to surface and subterranean waters alike. As used here, *hydrogeology* is not synonymous with *ground-water hydrology,* but, following May, implies the relationship between the geologic framework and any part of the hydrologic system.

Finally, *ground-water geology* refers to the geologic aspects of ground-water hydrology. Increasingly common in the literature, but seldom defined, is *hydrostratigraphy*. Ideally, it pertains to the classification of traditional stratigraphic units on the basis of their behavior as aquifers, aquitards, or, less commonly, aquifuges. The resulting subdivisions are variously referred to as *hydrostratigraphic units, hydrogeologic units,* or *hydrologic units*. (These terms are discussed in more detail in Chapter 4.)

Those new to the field will encounter other unfamiliar words and concepts in hydrogeologic reports. New professionals are encouraged to (1) check and learn the correct meaning of new terms as they come upon them, (2) practice using them appropriately, and (3) insist that others do so as well. Glossaries are readily available now, even on the Internet. However, because terms on the Net have not necessarily been subjected to technical review, they can contain erroneous definitions. For example, one such Internet glossary of hydrologic terms erroneously defines artesian wells as those in which water flows at the surface (water need only rise above the level at which it was encountered for the well to be artesian). Obviously, such sources should be used with caution.

REFERENCES

Gray, D. M. (ed.) 1973. *Handbook on the principles of hydrology.* New York: Water Information Center.

Howell, J. V. chairman. 1962. Glossary of geology and related sciences. Washington, DC: American Geological Institute. 325 p.

May, J. P. 1976. Geohydrology/hydrogeology: Letter to the editor, *Geotimes,* May, p. 15.

Meinzer, O. E. 1923. The occurrence of ground water in the United States—with a discussion of principles. Water-supply paper 489. U. S. Geological Survey. 321 p.

Todd, D. K. 1980. *Groundwater hydrology.* New York: John Wiley & Sons. 535 p.

Ward, R. C. 1967. *Principles of hydrology.* New York: McGraw-Hill. 403 p.

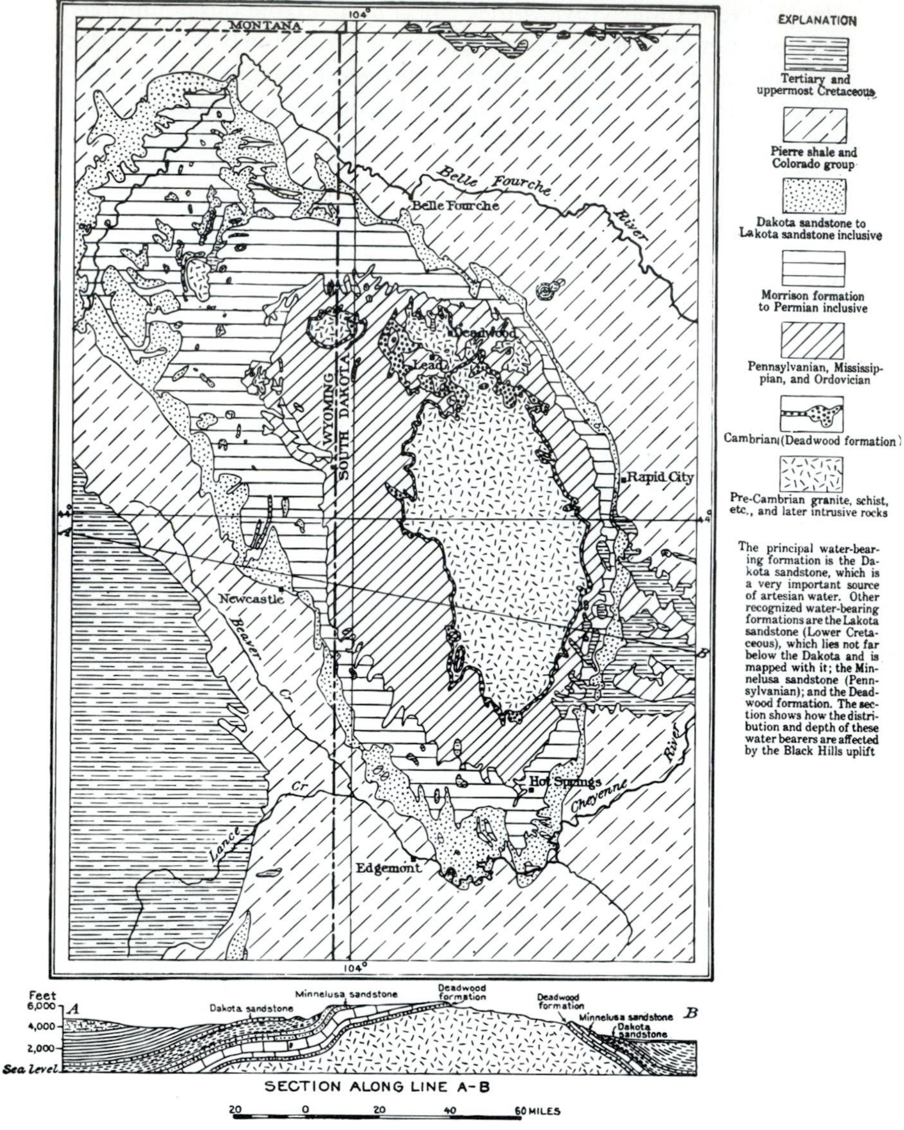

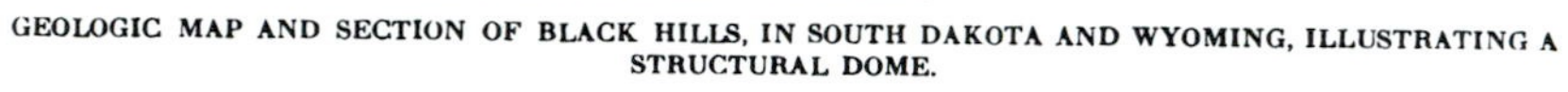

GEOLOGIC MAP AND SECTION OF BLACK HILLS, IN SOUTH DAKOTA AND WYOMING, ILLUSTRATING A STRUCTURAL DOME.

After N. H. Darton.

PART I

THE GEOLOGIC SETTING

Since ground water is intimately tied to the natural material within which it occurs, hydrogeologists must have a clear understanding of the geologic framework associated with a study area. This understanding comes from compiling available geologic information, characterizing the geologic setting, and evaluating the performance of geologic materials present as aquifers.

CHAPTER 2

Compiling Geologic Information

Since a sound hydrogeologic study requires a thorough understanding of the geologic framework within which the hydrologic system operates, a good way to start a project is to compile available geologic information. This process requires a knowledge of the geologic information needed, the regional geologic province involved, the usual sources of geologic data, and useful tools for data compilation.

GEOLOGIC INFORMATION

It is important to consider what kinds of data are needed before undertaking any research. Generally, the focus of the literature search is anything that will help characterize the geologic setting. The information sought will fall into one of three categories: stratigraphy, structure, and geomorphology.

Stratigraphic Information

Of particular interest is information on the geologic column for the area, that is, the stratigraphic units present and their basic properties. Specifically this includes

formal nomenclature
kind of material
thickness
areal extent
texture
composition
age

Variation in these properties over the region or study area are also of interest. For example, does a particular unit pinch out or change texture within the area of interest? Also, do any units intertongue in the area?

Structural Information

The subsequent modification of the strata by tectonic processes is also important. Thus, information is needed concerning any

folding
fracturing
faulting

Details about such things as location, type, size, and orientation of structural features are critical. Also, if the study area lies within or contains a major structural feature (e.g., Michigan Basin, San Rafael Swell, Rio Grande Rift, etc.), provide that information.

Geomorphologic Information

Landforms may also play a role in the hydrologic system. The result of the literature search should, therefore, include a compilation of available information on the major

erosional features
depositional features
petrogenic features
soils

Information about the location and characteristics of the major geomorphic features, as well as the affected or associated strata, should be compiled. Soils (as distinguished from unconsolidated sediments) may also be important, especially in the unsaturated zone. Thus, the major soil types and their characteristics across the area should also be researched.

Data Sheets

To facilitate the compilation process, develop data sheets employing the parameters given previously. A separate sheet may be necessary for each of the three major categories of information. As you gain experience with the data sheets, you may find that some items should be added and others revised or deleted. Once refined, such forms can be photocopied for routine use.

GEOLOGIC PROVINCES

Regional differences in landforms across continents, and the fact that they reflect differences in geology, have long been recognized (Thornbury 1965). The training of every geologist in the United States includes an awareness of these so-called "physiographic provinces" (Figure 2-1). Within the boundaries of these provinces, landforms are indeed similar. Since stratigraphy and structure are also more or less uniform within such regions, we may think of them as geologic provinces as well.

The existence and recognition of such provinces is of considerable use to a hydrogeologist. If you know in what province a study area lies, then you also know, in at least a general way, the geologic conditions of the area. If no geologic information is available for the study area, a report on an adjacent area in the same province will be helpful. This is of particular use in foreseeing drilling conditions, anticipating hydrostratigraphic units, recognizing geologic controls on the hydrologic systems, etc.

1. Superior Upland
2. Continental Shelf
3. Coastal Plain
4. Piedmont Province
5. Blue Ridge Province
6. Valley and Ridge Province
7. St. Lawrence Valley
8. Appalachian Plateaus
9. New England Province
10. Adirondack Province
11. Interior Low Plateaus
12. Central Lowland
13. Great Plains Region
14. Ozark Plateaus
15. Ouchita Province
16. Southern Rocky Mountains
17. Wyoming Basin
18. Middle Rocky Mountains
19. Northern Rocky Mountains
20. Columbia Plateau
21. Colorado Plateau
22. Basin and Range Province
23. Cascade-Sierra Nevada Mountains
24. Pacific Border Province

FIGURE 2-1 Geologic provinces in the United States. (Modified from U.S. Geological Survey, 1969.)

Different workers have distinguished different numbers of provinces. In the first attempt to subdivide the entire United States into provinces, Powell (1895) recognized 28 regions. Thornbury (1965) described 23 provinces in the conterminous United States. Shimer (1972) was content with a mere 19 provinces, and a U. S. Geological Survey (USGS) flyer shows only 13 (Figure 2-1).

Although it is beyond the scope of this book to describe the geologic provinces in detail, Table 2-1 gives a brief summary of their basic characteristics. Notice how they differ or what is unique about each province. More detailed information is available in regional geomorphology books. Thornbury (1965) is a good starting place, since that book gives references and an overview of each major province.

SOURCES OF GEOLOGIC INFORMATION

Once the geologic province and regional geologic conditions are determined, it is time to compile detailed local information. A little effort spent researching existing data at the outset will save time, effort, and expense later on. But where do you look? There is no need to feel lost when working in an area for the first time. In most states, and countries for that matter, you will find that government agencies are excellent sources of geologic information. In the United States, geologic information is available from several

TABLE 2-1 Brief Summary of the Geologic Provinces in the Conterminous United States

Province	Major Rock Type	Major Structure
Pacific Border	Sedimentary/volcanic	Uplifts, synclines
Sierra Nevada	Granitic intrusions	Metamorphic with faulting
Cascade Mountains	Explosive volcanics	Regional uplift
Basin and Range	Igneous, sedimentary, metamorphic	Block faulting, folding, thrusting
Columbia Plateau	Basalt flows, sedimentary interbeds	Horizontal except mountains
Colorado Plateau	Sedimentary, igneous	Gentle dips, monoclines, salt anticlines
Rocky Mountains	Igneous, sedimentary, metamorphic	Uplifts
Great Plains	Sedimentary	Horizontal except domes and basins
Interior Highlands	Sedimentary	Domes, strong folding/faulting
Central Lowlands	Sedimentary, glacial deposits	Major domes and basins
Superior Upland	Igneous, metamorphic	Very complex
Coastal Plain	Unconsolidated sediments	Gentle dips, local folding/faulting, salt domes
Piedmont	Igneous, sedimentary, metamorphic	Complex, fault troughs
Blue Ridge	Igneous, metamorphic	Thrusting
Ridge and Valley	Sedimentary	Folding, faulting
Appalachian Plateau	Sedimentary	Regional syncline
Interior Low Plateau	Sedimentary	Broad features, local folding/faulting
Adirondack Mountains	Igneous, sedimentary, metamorphic	Anticlinal
New England	Igneous, metamorphic	Continuation of Piedmont, Blue Ridge, Ridge and Valley structure

*See Figure 2-1 for locations.

Source: Compiled from Thornbury (1965).

sources. These mainly include state and federal government agencies, but universities, professional organizations, and various periodicals may prove useful as well.

Government agencies are not only outlets for printed material, they can also help you find the specific information needed. Both state and federal agencies prepare bibliographies of geologic works or selected topics, regardless of source. Some agencies have computerized databases as well.

State Geological Survey

In both the United States and abroad, the State Geological Survey (also called by other names, such as State Bureau of Geology, State Bureau of Mines and Mineral Resources, State Bureau of Economic Geology, etc.) is the best source of information on that state. When beginning to work in a state, it is a good idea to locate the geological survey and get a copy of its publication list. Check to see if there is a separate list for open-file reports. Such reports, though not subjected to the same level of editorial scrutiny or peer review as formal publications, contain valuable information. In addition to publishing reports and maps on a wide range of subjects, lexicons of stratigraphic names, and bibliographies, state surveys sometimes maintain a library of logs, cuttings, and core from oil and gas wells, mineral exploration holes, and even water wells, as well as archives of old mine records.

In some countries, state geological surveys predate and have primacy over the national geological survey. This is the case in Australia, for example, where the national Bureau of Mineral Resources (BMR) was not formed until well after the state geological surveys were in place. As a result, the national survey usually becomes involved only in studies spanning two or more states.

State surveys are also an important source of illustrations. Some have a very helpful policy regarding their published illustrations: All are numbered and filed so they can be readily retrieved. Thus, if you come across a map or diagram in one of their reports that would be useful in your work, it is possible, using the reference number, to get a hard copy or slide made. This is the case, for example, with the South Australia Department of Minerals and Energy. If you wish to use an illustration appearing in a state survey publication in your area, check to see if they have a similar policy.

National Geological Survey

The national geological survey of a country is also a valuable source of information. In the United States, the U.S. Geological Survey is the main source of geologic information about an area, after the state survey. It not only produces topographic and geologic maps, but also lexicons, bibliographies, and publications dealing with various geologic topics. This agency maintains an office in a major city, if not the capital, in all states. The State Geological Survey usually carries and distributes the topographic maps and pertinent USGS geologic maps and reports on that state. If not, such material can be obtained through the district (state) or regional USGS office.

Outside the United States, a national geological survey is a similar source of information. In Australia, for example, the BMR produces regional and national reports, as well as a geologic atlas for the country. As previously discussed, they lead in regional studies spanning two or more states and are thus the source of information on larger geologic features, such as the Great Artesian Basin and the Murray Basin.

Universities

Universities in the study area's state are also a good source of geologic information. Libraries of universities that have graduate geoscience programs will have copies of any geologic theses or dissertations done there. They may also have copies of theses and dissertations written by students at other universities within the state, and may have a bibliography of geologic theses and dissertations done in the state.

Some universities also publish their own periodicals containing faculty research. A few of these publications focus strictly on geologic work; others at least include it. These sources may be the only place some studies are reported, and they should not be overlooked.

Note that outside the United States there are often two categories of universities: traditional (older, offering liberal arts subjects) and applied (newer, offering engineering and technical subjects). The geology department and associated library holdings will usually be found on the campus of the older, more traditional university.

Professional Organizations

Many states or regions have active geological societies that you should include as additional sources of information. They, too, may have prepared lexicons for their members. Of special interest, however, are their guidebooks and special reports, which often contain information not available anywhere else. The New Mexico Geological Society, for example, has over the years produced a comprehensive set of annual field conference guidebooks covering the major geologic regions of the state. Because it is their policy to keep these in print, even the earliest guidebooks are still available. Proceedings of a state's academy of science may also contain geologic papers.

Journals published by national geological organizations may also contain papers on the study region or area. The number of such organizations is rapidly increasing, but they include such groups as the Geological Society of America, American Association of Petroleum Geologists, and Sigma Gamma Epsilon, to name a few. Some professional organizations, though not restricted to geology, publish geologic papers (for example, the American Association for the Advancement of Science and Sigma Xi).

A number of journals are not affiliated with any professional organization. Several of these are devoted to geology and may contain pertinent information (for example, *Journal of Geology*).

Internet

The Internet is the newest source of geologic information. Most of the sources previously listed have a home page. Less formal sources also have sites. These sites range from national parks to "friends" of a given geologic time interval or topic. The search for geologic information is no longer complete without a trip to the net.

GEOLOGIC TOOLS

According to the dictionary, a tool is anything necessary for a person to conduct a trade—a means to an end. Geologic tools thus include a host of field and laboratory instruments, as well as numerous types of illustrations. This book assumes that readers are more familiar with the common devices used in field or laboratory geologic studies (Brunton compass, Jacob staff, hand level, microscope, etc.) than they are with the possible means of presenting the results of such studies. As used here, then, "geologic tools" are the various types of data and illustrations that may be used first to compile and eventually to characterize the geology of an area. For discussion purposes, these may be grouped into surface, subsurface, and laboratory tools, on the basis of their data source.

Surface Geologic Tools

Considerable information about the geologic framework can be gained from observations made at the earth's surface or by using surface data. Such observations may be made from topographic maps, geologic maps, aerial photographs, and outcrops.

Topographic maps may be used to learn about the general geologic structure and something about the rock type or deposits in the area. For example, drainage patterns, clearly shown by topographic maps, suggest the geologic structure and material at the surface (Figure 2-2). The presence of karst features, such as depressions (sinkholes) accompanied by disappearing streams, indicates an abundance of carbonate or evaporite strata. Poorly integrated drainage and water-filled depressions (kettle lakes) attest to the presence of continental glacial deposits at the surface.

As might be expected, geologic maps give the most information about the geologic framework. The legend accompanying such maps constitutes a geologic column for the area. Even the brief descriptions of rocks and sediments given in the legend will help a hydrogeologist decide what units may be significant hydrologically. The location and type of faults in the area will also be shown. These are important because they may play a role in controlling water movement.

Aerial photographs are an excellent supplement to both topographic and geologic maps. Air photos not only show the general relief or lay of the land, but show such useful things as road conditions, ground cover and new development (at the time the photo was taken). They may also be used as a base for mapping field observations on geology, soils, vegetation, land use, etc.

Outcrops are the ultimate source of geologic information. Combining topographic maps, geologic maps, and air photos, possible outcrop areas of key stratigraphic units may be identified. Visit these sites for a first-hand examination of the major geologic materials in the study area. If no detailed description of the geologic units exists in the literature on the area, it may be necessary to measure and describe some stratigraphic sections. These procedures can be found in field geology or stratigraphy texts. Comparing observations from several sites will show the lateral and vertical homogeneity of the units. Such information may provide a basis for accepting or rejecting a given conceptual geologic model for the area. Field work may also include sampling major geologic materials for laboratory analysis.

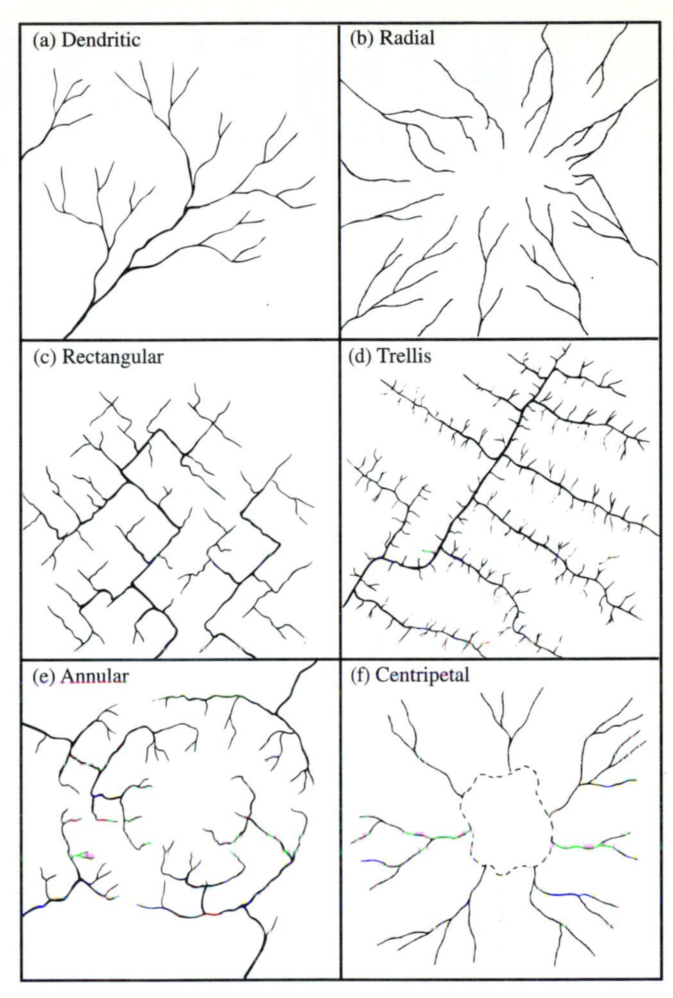

FIGURE 2-2 Drainage patterns reveal the geologic make-up, history, or structure of the region: (a) horizontal, granular strata; (b) elevated area, such as intrusion or volcano; (c) jointed rock; (d) folded or tilted strata; (e) breached dome; (f) closed depression of structural, volcanic, or eolian origin. (From *The Origin of Landscapes* by H. F. Garner, Figures 2.19 and 2.20. Copyright © 1974 by Oxford University Press, Inc. Used by permission of Oxford University Press, Inc.)

Subsurface Geologic Tools

To completely characterize the geologic setting of an area, you will need some observations from the subsurface. Such information may already be compiled into various types of subsurface-geology maps and geologic cross sections. If these sources are not available, basic well data may be used.

Well data include various types of information:

observations made by the driller
descriptions of cuttings or core by geologists
geophysical logs made
depth and elevation of formation tops
intervals sampled or cored
samples (cuttings or core) preserved from the hole

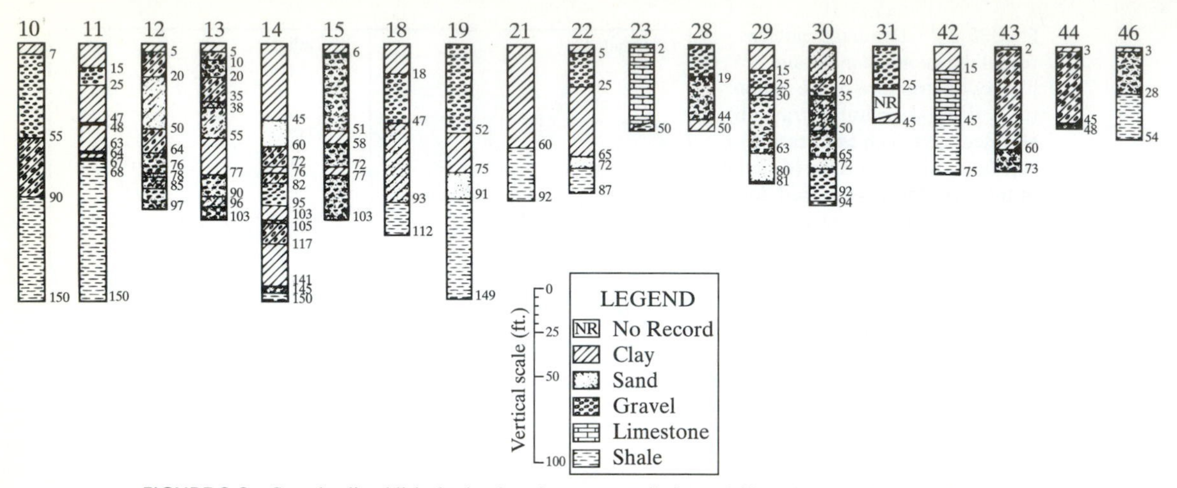

FIGURE 2-3 Standardized lithologic plots for water wells from driller's logs, as done by the state of Ohio (From Groundwater Resource Evaluation by W. C. Walton, Figure 2.24. Copyright 1970 by McGraw-Hill. Reproduced with permission of The McGraw-Hill Companies.)

Various government agencies keep such information and increasingly are making computerized databases to help locate the observations needed. Some states plot simple lithologic logs from the driller's logs for all water wells and provide these to the public (Figure 2-3).

Drillers are not normally trained in geology and their logs are notorious for geologic inaccuracies. Unfortunately, that information may be all that is available. However, by comparing logs made for the same hole by both the driller and by the geologist, you can develop a basis for obtaining useful geologic information where only the driller's log is available. This, of course, is only possible when drillers are consistent in their use of terminology.

Bredehoeft and Farvolden (1963) successfully applied this approach when assessing the water-resource potential of alluvial materials in northern Nevada. Lacking funds for test drilling, they had to rely on existing data, largely driller's logs. They standardized interpretations of driller's lithologic descriptions, as shown in Table 2-2. The main aquifer was found to be material described by drillers as "gravel." Lithofacies maps (for gravel thickness and sand/shale ratio) were made for selected depth intervals, using the standardized data. Statistical analyses showed a significant relationship existed between specific capacities (yield/ft of drawdown) calculated from crude pumping tests and the driller's lithologic descriptions.

The most geologically significant aspects of a driller's log are usually the observations as to changes in drillability. Major formation boundaries may be located in this way. For example, the alluvium/bedrock contact is usually easily picked by the driller. However, if the bedrock is young and relatively soft or poorly consolidated sedimentary material, an anomalously thick alluvial sequence may be reported.

Consider this case in northwestern New Mexico. The thickness of Quaternary valley-fill alluvium in the region was generally found to be 100 ft or less. However, a driller reported atypical thicknesses of 170 and 308 ft for the alluvium at two sites in the Animas River Valley. Since the alluvium is the main aquifer in that area, these unusual thicknesses were of considerable interest. The depth to bedrock was checked by making

TABLE 2-2 Geologic Interpretation of the Drillers' Description of Sediments

Driller's Description	Geologic Interpretation	Percent Gravel[a]
Gravel	Gravel	100
Cement gravel	Gravel, pebble-sized grains predominate	100
Sand and gravel	Interbedded beds of medium to coarse-grained sand with beds of gravel	50
Gravel and clay	Pebbles and larger clastic material in a matrix of fine sand and silt; interbedded with some beds of gravel. (Probably mudflow deposits with some interbedded stream sediments)	0–25
Sand	Sand, medium to coarse grains	0
Sandy clay	Interbedded, clay, silt, and fine- to medium-grained sand	0
Silt clay	Silt with minor amounts of clay	0
Yellow clay	Interbedded clay, silt and fine-grained sand (possibly, at least in part, lacustrine)	0
Blue clay	Clay, blue, thinly bedded (probably lacustrine)	0
Lava rock	Either volcanic flows or volcanic detrital material	?

[a]Gravel is used to describe a clastic deposit in which the median grain size is 2 mm. or larger with a matrix of predominantly medium to coarse grained sand.

Source: Bredehoeft and Farvolden, 1963.

soundings with a small, portable seismic instrument (sledge-hammer-and-plate type). This showed the alluvium to be less than 100 ft at both sites and, thus, not unusual for the region (Brown and Stone 1979). Apparently, the soft nature of the Tertiary strata beneath the valley-fill made it difficult to detect the contact during drilling.

Even geologist's logs vary in quality and usefulness. There may be a good reason for differences in the amount of detail between logs, but differences within the same log are inexcusable. It is most disconcerting, for example, to try to use a log that concentrates on color for one sample, while making no mention of texture, and then goes on and on about texture for the next sample, but omits anything about color. A standard list of parameters for cuttings or core description should be adopted and faithfully employed for each sample. Some suggested items to include are given in Appendix A.

Many characteristics of subsurface materials can be learned from geophysical logs (Figure 2-4). These are continuous records of mechanical, spontaneous, or induced measurements of formation parameters, made with a tool lowered down the open borehole on a wireline (Table 2-3). Variation in hole diameter can be determined from a mechanically produced caliper log. Variations in borehole (often water) temperature, electrical current, or natural radioactivity may be learned from spontaneously measured temperature, self-potential (SP), and gamma-ray logs, respectively. Resistance to electrical current, velocity of sound propagation, and reaction to gamma-ray or neutron bombardment is depicted by induced measurements on resistivity, induction, sonic, density, and neutron logs, respectively. Various books (for example, Asquith and Gibson 1982; or Rider 1986), as well as service-company manuals, give detailed instructions on the use and interpretation of geophysical logs.

"Scout cards," published by the American Petroleum Institute, give information on oil and gas wells and are another good source of subsurface data (Figure 2-5). These are often filed by a state agency: geological survey, or oil and gas commission. The cards routinely give such information as well location, ground-surface elevation, total well

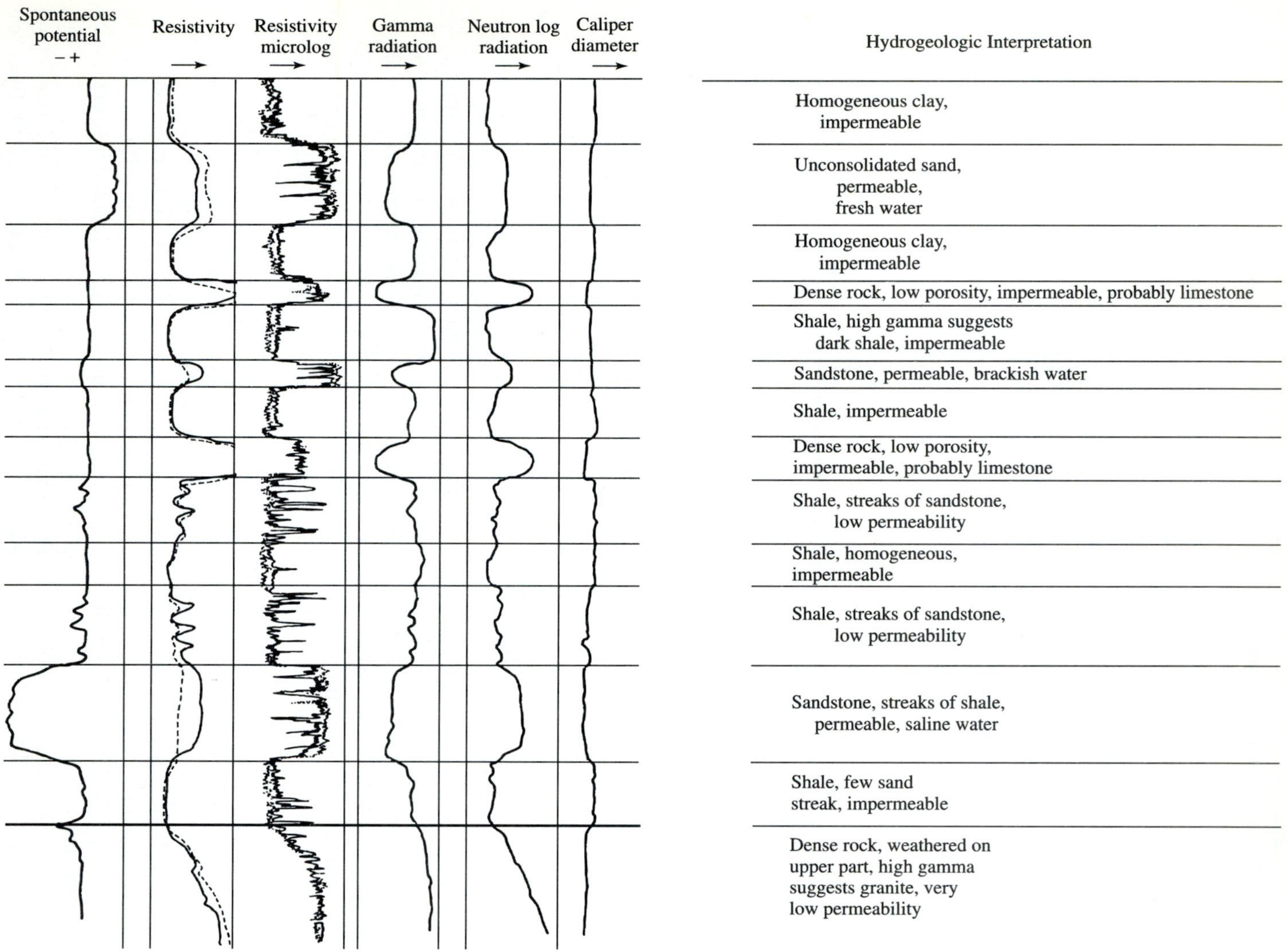

FIGURE 2-4 Hypothetical sequence of sedimentary strata and typical signature on major geophysical log types. (Reproduced from Davis and DeWeist 1966, Figure 8.33 with permission of the authors.)

TABLE 2-3 Summary of Wireline Geophysical Well Logging

Measurement Type	Log Type	Parameter Measured
Mechanical	Caliper	Borehole diameter
Spontaneous	Temperature	Borehole temperature
	SP (self-potential)	Spontaneous electrical currents
	Gamma-ray	Natural radioactivity
Induced	Resistivity	Resistance to electrical current
	Induction	Conductivity of electrical current
	Sonic	Velocity of sound propagation
	Density	Reaction to gamma-ray bombardment
	Neutron	Reaction to neutron bombardment

Source: Modified from Rider (1986).

depth, and the depth and elevation of major formation tops. They also indicate what intervals were sampled or cored and the types of logs made. Sample portions or "cuts" are often archived, and copies of logs are often filed with the state geological survey or oil and gas commission.

Maps of various subsurface conditions may exist for selected formations. Most common are the structure map (showing the elevation of the top of a unit), the depth map (showing the depth to the top of a unit), and the isopach map (showing either the drilled or true thickness of a unit). It takes all three maps to minimally characterize subsurface strata. This basic suite of maps was compiled for each sandstone aquifer in a hydrogeologic study of the San Juan Basin in northwest New Mexico (Figure 2-6). Other types of subsurface maps may show such things as the ratio, cumulative thickness, and percent of given grain sizes in clastic sediments or sedimentary rocks. More specifically, maps may be prepared to show sandstone/shale ratio (based on thickness), total sandstone thickness, percent sandstone (by thickness), sandstone grain size (Figure 2-7, p. 25), and extent of sandstone bodies (Figure 2-8, p. 26). If such maps do not already exist for the study area, and the data are available, it may be useful to make some. Petroleum geology and stratigraphy texts give information on map construction (e.g., Low 1958; Prothero 1990, chapter 9).

Cross sections are not only the quickest way to learn the geologic conditions beneath an area, but are also indispensable for depicting geology in a report. Such illustrations show the thickness, depth, and extent of stratigraphic units, as well as their deformation by folding and faulting (Figure 2-9, p. 27). They may be drawn using any combination of drill-hole data, outcrop observations, and geophysical information (seismic, gravity, etc.). Construction may require using different vertical and horizontal scales in order to clearly show relationships in the space available. This is acceptable, as long as the scales are indicated or vertical exaggeration is clearly stated (vertical exaggeration = vertical scale/horizontal scale, for example, 1″/500′ / 1″/2,000′ = 4 times or 4×). Ideally, in such cases, a cross section with no exaggeration is also presented for comparison.

Maps and sections depicting geophysical parameters are informative and may also be used to compile the area geology. These items might include gravity maps and seismic sections. O'Brien and Stone (1984) found that various types of geophysical data

Amerada Petr. Corp. 1 Anderson, S.E. "A" 19-9S-35E

COMPANY WELL NO. LEASE S T R

Loc. 660 fr. S L 660 fr. E L County Lea

Spud. 5-23-63 Comp. 8-26-63 Field Undes.

T.D. 12825 P.B.11.500 T.Pay

I.P. CAOF 4600 MCFGPD. SI TP 391#

Remarks: Prod. Int. 11477-484 (Atoka), Cored 9640-9700, rec 60 being:16 dense ls., 6-1/2 sh. 2-1/2 dense ls., 7 ls. good vuggy to PPP. NS. 26, dense ls. 2 ls., very good vuggy to PPP.NS, Cored 9700-08, rec 8 being: 1 ls, very good vuggy & PPP. NS 3-1/2 dense ls. 1/2 very vuggy lo. NS. 1 dense ls 2 sh, DST 9627-9708. oo 1 hr. rec 120 DM, NS, FP 44-107#, 90 FSIP 3439#. Cored 9708-50 roc 42 being 3 sh, 2 ls, horiz. fracs., 3-1/2 dense shly lo. 2-1/2 sh, 1 shy ls, 2-1/2 dense ls, 3-1/2 sh, 5xxxxxxxxx 4-1/2 dense ls, 2-1/2 ls, fair-good vuggy PPP, NS, 6 ls, very good vuggy & PPP, NS, 11 dense ls, horiz. hairline fracs. DST 9718-9750 op. 3 hrs, rec 6122 salt wtr. 180 salt wtr. CDM.

CSG. RECORD

Size	Depth	Scx
13-3/8	377	500
9-5/8	4030	1500
5-1/2	11607	500

Thg. at

FORMATION RECORD

Elev. 4182 DF

T.	by clectric log:	
T.	Yates	2762
T.	Queen	3466
T.	San And.	4035
T.	Glor.	5451
T.	Tubb	6901
T.	Abo	7735
T.	Bough "C"	9724
T.	(Over)	

FP 91-2669#, 90 FSIP 3143#. Cored 9750-8910, rec 60 being. 2 dense ls 6-1/2 sh, 23-1/2 dense ls, 13 shly ls, 8 sh, 6 dense ls, 1 ls, fair PPP, NS. Cored 9810-70, rec 60 being: 4-1/2 ls. good vuggy & PPP, NS, 28 dense ls, 1-1/2 sh, 7 ls, fair PPP, BO&G, 19 dense ls. DST 9820-70, open 1 hr, rec 5 gals DM, NS oil, gas or wtr, FP 61-61#, 90 FSIP 85# Cored 9870-9907, rec 37 bding: 6 sh, 31 shly ls, NS, DST 12628-660 op 1 hr, usec rcc 2800 WE + 30 DM, NS, FP 38-404#, 60 FSIP 1101#. DST 12600-695, op 1 hr, Used & Rcc 2800 WB, 45, DM, NS, FP 1361-1368#, 60 FSIP 4736#. DST 12694, 825, op 2 hrs, Used & rec 2800 WB, 2914 salt wtr & 564 salt wtr cut DM, FP 1445-2953#, 66 FSIP 4894#. Straddle pkr DST 1148-500 (Atoka), op 3 hrs, Used 1500 WB, WBTS in 10, G & Dist. TS, fld. to pts foor 45 mins, fld. 23 B Dist, in 2 hrs thru 1/2 ch, grav 62 (Corr.), SFP increased from 1800 to 1820, Gas vol. increased from 4149 MCF to 4199 MCFGPD, rec 270 dist. + 120 salt wtr, FP 2777-3157#, 90 FSIP 3350#. STRADDLE PKR DST 9400-9440 (Upper Penn), op 1 hr 5 mins, rec 20 DM, FP 25-24#, 50 FSIP 117#. Perf 14/11477-484, Natural.

Formation Record Cont'd

Cisco 9800	Atoka 1183	Woodford 12560
Canyon 10318	Morrow 11674	Dev. 12605
Strawn 10683	Miss 11836	

FIGURE 2-5 Typical "scout card" giving information for an oil well in Lea County, New Mexico. (From the files of the New Mexico Bureau of Mines and Mineral Resources.)

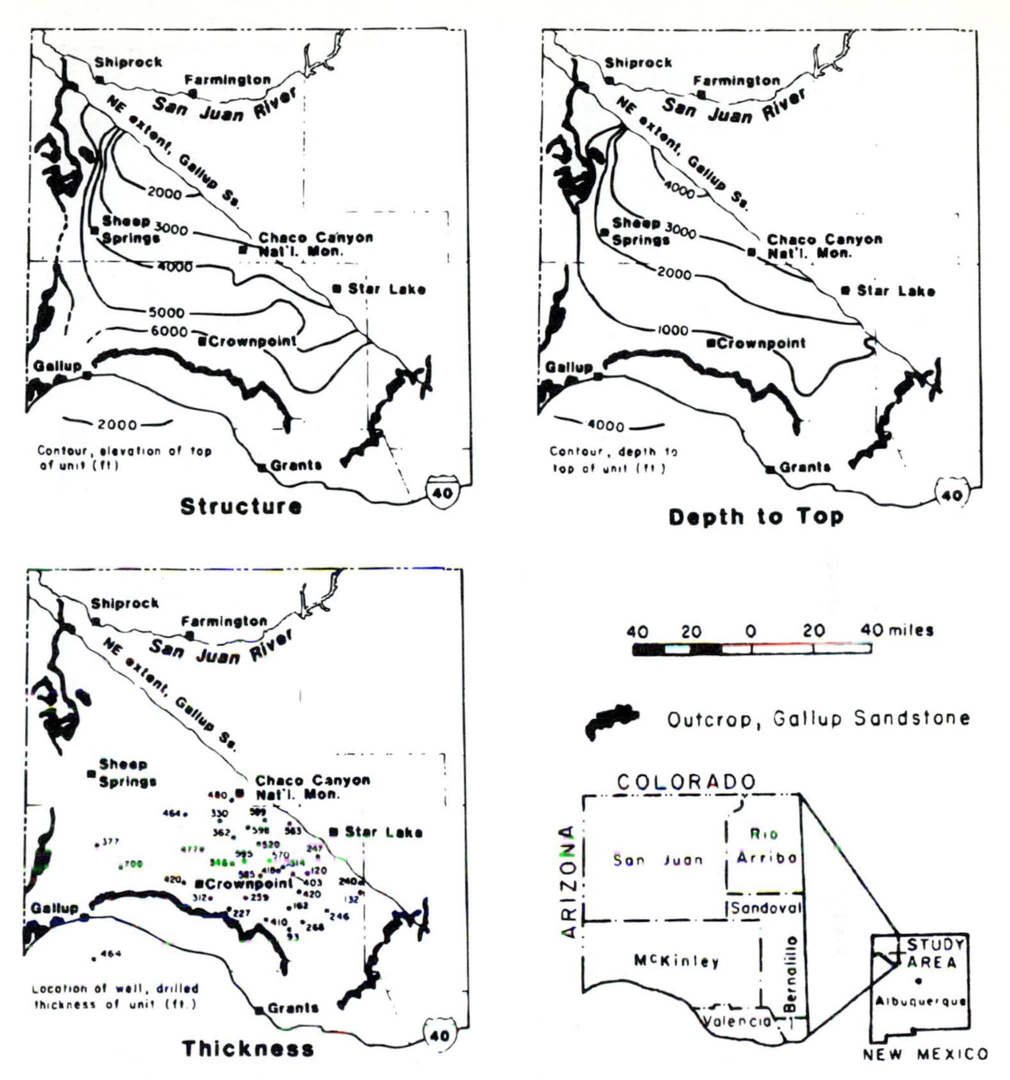

FIGURE 2-6 Suite of subsurface maps for the Gallup Sandstone aquifer, San Juan Basin, northwestern New Mexico. (From Stone 1981, Figure 3. Reprinted by permission of Ground Water Publishing Company.)

supplemented geologic information on an alluvial basin in southwestern New Mexico. More specifically, they used available complete Bouger gravity anomaly maps and seismic refraction profiles, together with driller's logs, to define variation in aquifer thickness (Figure 2-10, p. 28).

Laboratory Tools

Many characteristics of the study area's geologic materials can be determined only in the laboratory. Of special interest are texture and composition. Texture controls hydraulic properties and composition controls water quality. If data on these parameters are not

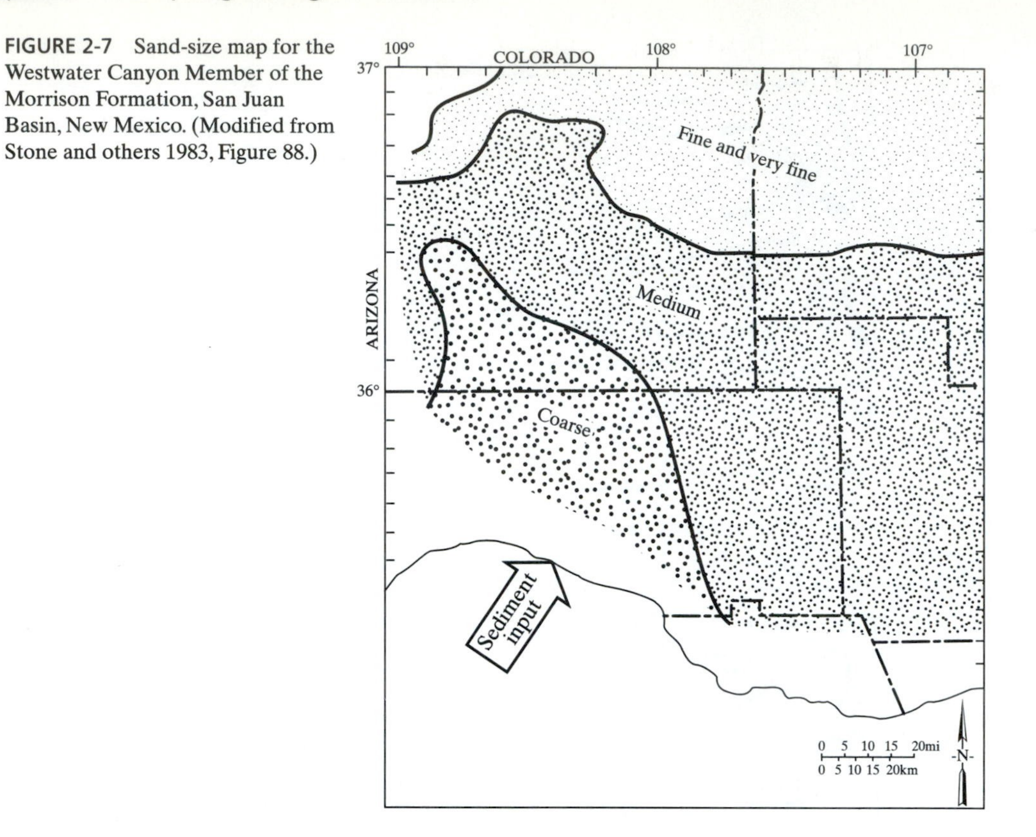

FIGURE 2-7 Sand-size map for the Westwater Canyon Member of the Morrison Formation, San Juan Basin, New Mexico. (Modified from Stone and others 1983, Figure 88.)

available for the major or potentially major hydrostratigraphic units, it may be useful to obtain some.

Texture is an important attribute of a geologic material. It includes the size, shape, and sorting of grains or crystals making up the geologic medium. The methods of studying these parameters vary with the type of material.

For unconsolidated or poorly indurated, fragmental (clastic) sedimentary material, grain size may be determined by sieving or by pipette or hydrometer analysis, as described in petrography texts (for example, Krumbein and Pettijohn 1938; Folk 1974). The most commonly occurring (modal) size is used to name the material: medium sand, fine gravel, etc. The range and distribution of grain sizes are used to define sorting (see Folk 1974). Grain shape is determined by examining the grains with a binocular microscope and comparing them to standard roundness and sphericity classes (for example those of Powers 1953). Commercially available comparators (cards with diagrams or actual grains representing the major sizes and shapes) are useful for field descriptions of texture.

In the case of well consolidated or indurated clastic material, nonclastic sedimentary rocks, and other crystalline rocks (igneous and metamorphic), textural characteristics may be determined through examination of a hand specimen with a binocular microscope, or of a thin section with a petrographic microscope. However, in some of these materials, texture is less important for determining hydraulic properties than are fractures or dissolution openings. Thus, such features should be noted during microscopic analysis.

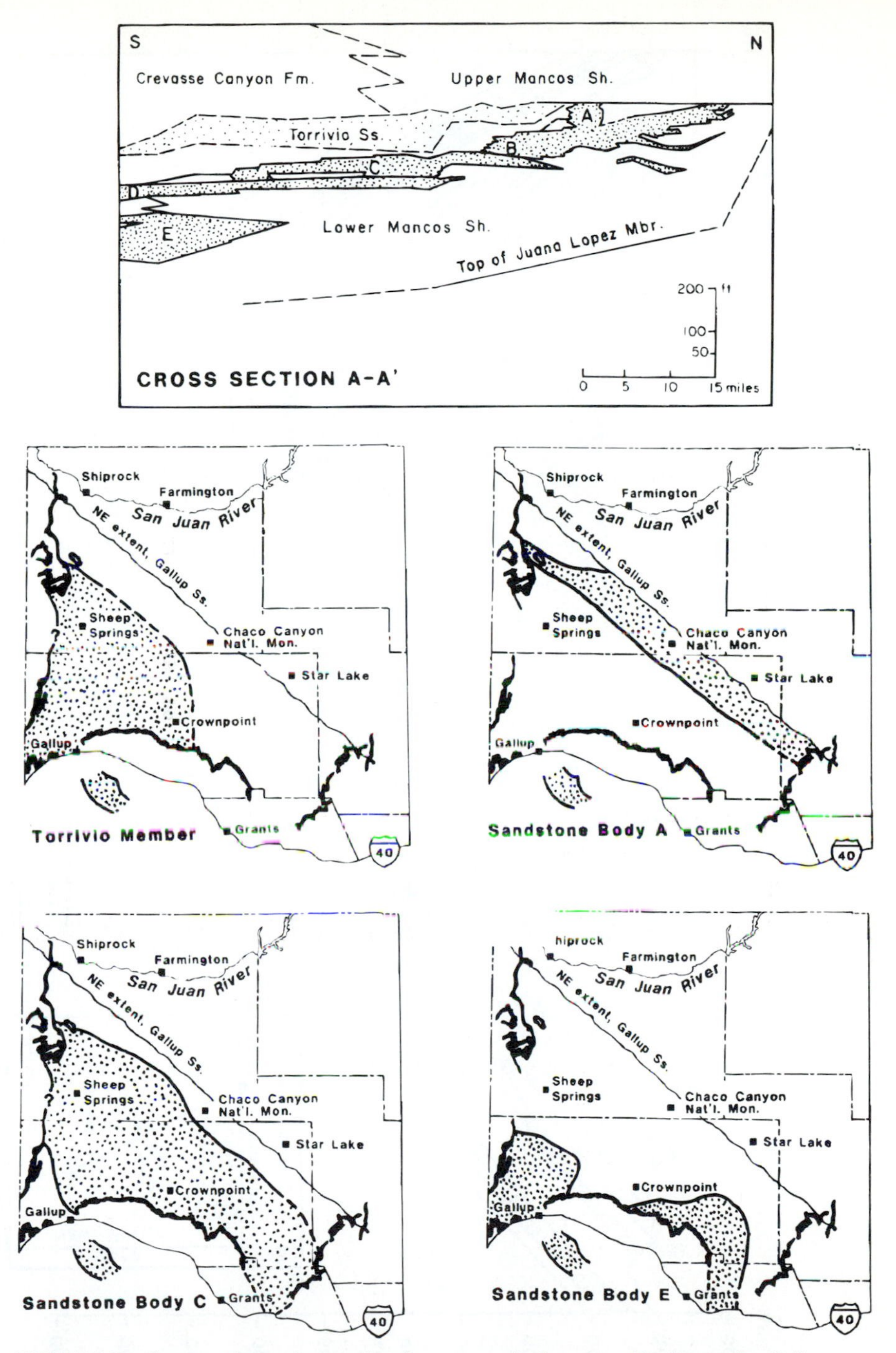

FIGURE 2-8 Cross section and maps showing extent of selected sand bodies in the Gallup Sandstone, San Juan Basin, New Mexico. (From Stone 1981, Figures 4 and 5.) See Figure 2-6 for location. Reprinted by permission of Ground Water Publishing.)

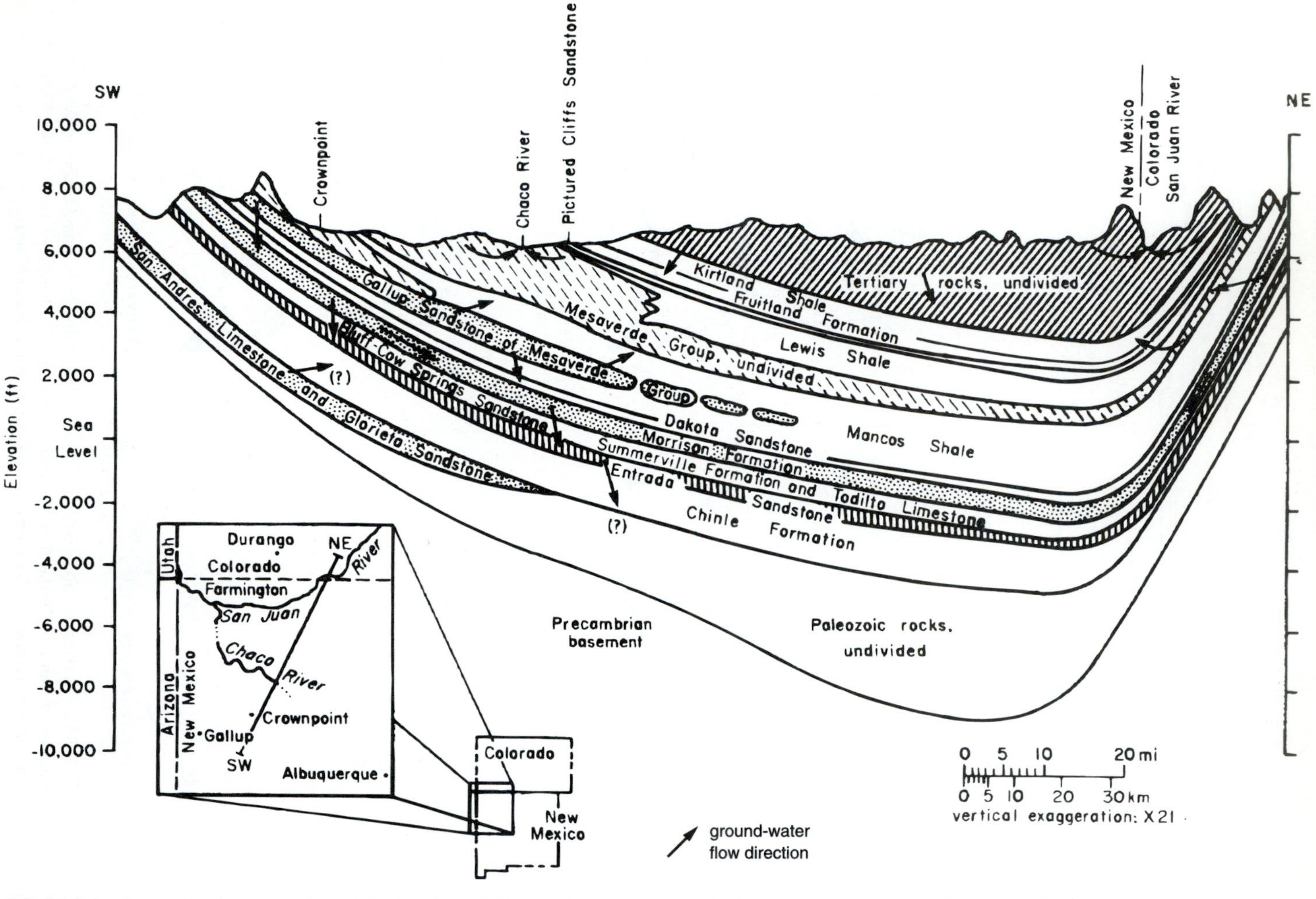

FIGURE 2-9 Generalized cross section of the San Juan Basin, northwest New Mexico. (From Stone and others 1983, Figure 10.)

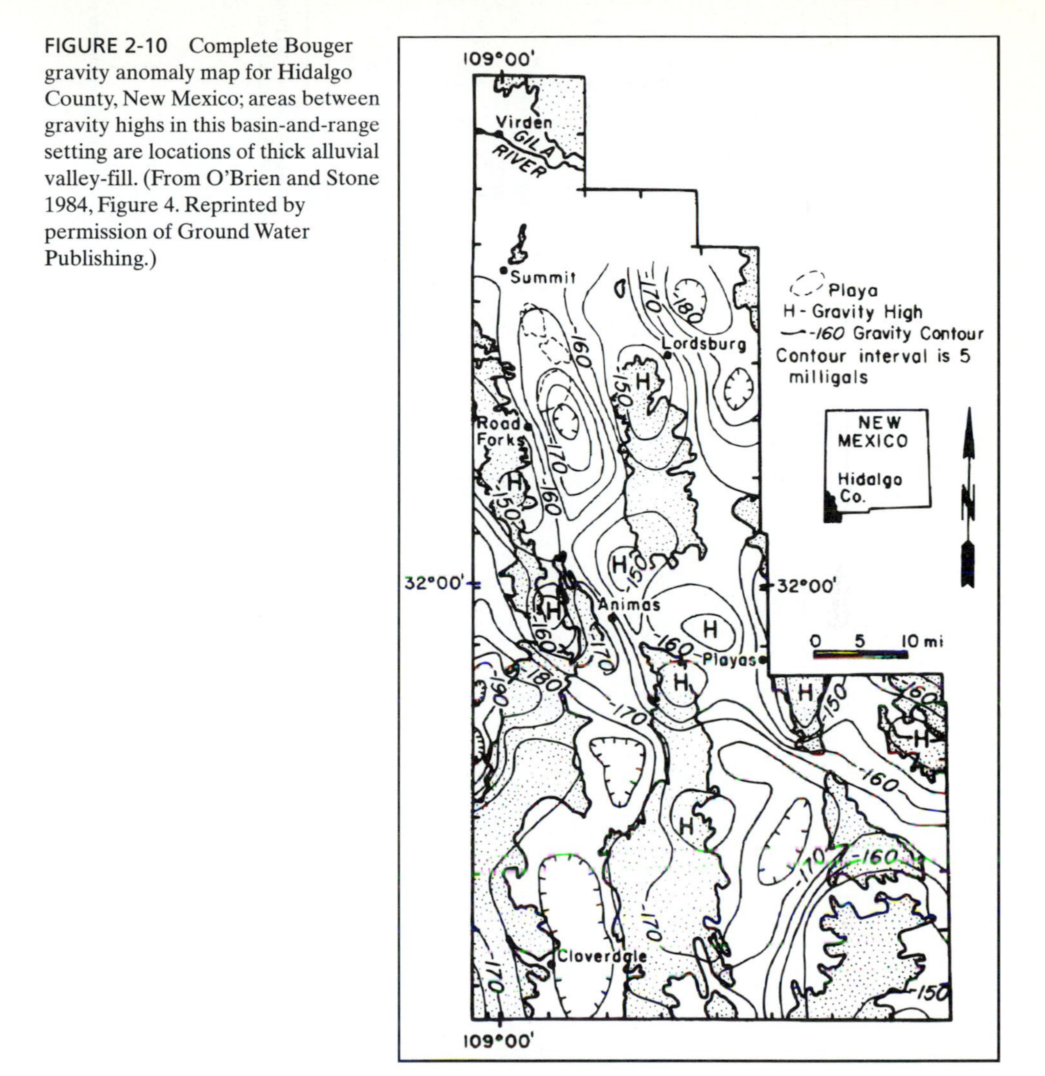

FIGURE 2-10 Complete Bouger gravity anomaly map for Hidalgo County, New Mexico; areas between gravity highs in this basin-and-range setting are locations of thick alluvial valley-fill. (From O'Brien and Stone 1984, Figure 4. Reprinted by permission of Ground Water Publishing.)

Characterizing the composition of a geologic material essentially amounts to describing the minerals present and their relative abundance. If the geologic materials in the study area have been previously analyzed and classified properly, their general mineral (and elemental) make-up can be assumed. If they have not been previously studied, and the reasons for water-quality variations in the area are not obvious, making such an analysis may be worthwhile.

Composition of a given material may be determined by several different methods, depending on its grain size and induration. Coarse grained materials may be examined in hand specimen by binocular microscope, or in thin section with a petrographic microscope. This may involve visual estimation of the abundance of mineral constituents or detailed point-counting (modal analysis). In the case of clastic materials, such as sandstones, point-counting should determine first the relative abundance of framework, ma-

TABLE 2-4 Example of Petrographic Data Compiled for a Hydrogeologic Study

Part I Textural Analyses																				
				Size fractions (φ)—% by weight							Mode		Median		Mean		Sorting			
Anal. No.	Unit	A/S	Sample No.	Location (¼, sec, T,R)	C <1	M 1–2	F 2–3	VF 3–4	S/C >4	Total	φ	Verb.	φ	Verb.	φ	Verb.	S_{ig}	S_g	Verb	S/M Ratio
19	Kg	LF	SJ3-5-3	NW,21,30N,19W	2.00	79.10	13.30	2.70	2.90	100.00	1–2	M	1.70	M	1.77	M	0.70		MW	33.5
20	Kg	LF	SJ3-6-1	2NW,17,27N,19W	0.30	20.40	64.40	9.80	4.90	100.00	2–3	F	2.40	F	2.60	F	0.91		M	19.4
21	Kg	LF	SJ3-6-3	NW,17,27N,19W	0.20	8.00	71.20	13.80	6.80	100.00	2–3	F	2.60	F	2.69	F		0.54	M	13.7
22	Kg	LF	SJ3-6-7b	NW,17,27N,19W	0.53	27.60	62.80	4.60	4.50	100.00	2–3	F	2.30	F	2.33	F	0.99		M	21.2
23	Kd	LF	SJ3-2-3	NO,26,30N,20W	0.02	19.40	70.70	4.70	5.20	100.02	2–3	F	2.40	F	2.41	F	0.79		M	18.2
24	Kd	LF	SJ3-3-10	SE,34,28N,20W	1.30	76.00	14.50	2.90	5.30	100.00	1–2	M	1.70	M	1.83	M	0.92		M	17.9
25	Kd	BB	BB-6	SE,14,13N,10W	0.05	6.15	49.03	30.74	14.04	100.01	2–3	F	2.50	F	2.67	F	0.91	0.75	M	6.1
26	Jmb	LF	SJ3-3-4a	SE,34,28N,20W	0.70	13.60	71.30	8.50	6.00	100.00	2–3	F	2.50	F	2.51	F		0.47	MW	15.7
27	Jmw	BB	BB-7	SE,14,13N,10W	91.60	2.75	3.10	1.59	0.98	100.02	<1	VC-G	<1.00	VC-G	1.00	VC-G	1.26	1.00	P	101.1
28	Jmw	BB	BB-8	SE,14,13N,10W	38.30	44.54	12.07	2.75	2.32	99.98	1–2	M	1.00	C	1.17	M	0.83	0.75	M	42.1
29	Jms	LF	SJ3-1-10b	SW,14,30N,21W	1.10	15.70	52.70	27.00	3.40	99.90	2–3	F	2.50	F	2.66	F	1.11		P	28.4
30	Jb	BB	BB-9	SE,14,13N,10W	0.24	31.69	50.41	10.00	7.68	100.02	2–3	M	2.00	M	2.17	F	1.36	0.75	P-M	12.0
31	Js	LF	SJ3-1-7	SW,14,30N,21W	0.00	0.60	18.50	72.30	8.50	99.90	3–4	VF	3.40	VF	3.38	VF		0.48	MW	10.8
32	Js	LF	SJ3-1-8	SW,14,30N,21W	0.70	1.40	43.50	41.40	13.00	100.00	2–3	F	3.10	VF	3.18	VF		0.67	M	6.7
33	Jeu	LF	SJ3-1-3	SW,14,30N,21W	0.00	4.80	47.00	38.90	9.30	100.00	2–3	F	3.00	F	3.05	F		0.68	M	9.8
34	Jeu	BB	BB-10	SW,23,13N,10W	0.16	11.01	62.49	16.35	9.19	99.20	2–3	F	2.50	F	2.50	F	1.24	0.50	P-M	9.8
35	Jem	LF	SJ3-1-2	SW,14,30N,21W	0.00	0.03	14.40	73.10	12.40	99.93	3–4	VF	3.50	VF	3.50	VF		0.45	MW	7.1
36	Jem	BB	BB-11	SW,23,13N,10W	0.00	0.72	3.67	52.04	43.56	99.99	3–4	VF	3.50	VF	>4.00	Slt	1.88	2.25	P-EP	1.3
37	Trw	BB	BB-12	SW,23,13N,10W	0.34	72.35	17.94	2.64	6.73	100.00	1–2	M	1.50	M	1.67	M	1.19	0.25	M	13.9

Anal. no = analysis number
Units:
Kk (f) = Farmington Sandstone Member of Kirtland Shale
Kpc = Pictured Cliffs Sandstone
Kch = Cliff House Sandstone
Kmf = Menefee Formation
Kpl = Point Lookout Sandstone
Kcda = Dalton Sandstone Member of Crevasse Canyon Formation
Kmm = Mulatto Tongue of Mancos Shale
Kg = Gallup Sandstone
Kd = Dakota Sandstone
Jmb = Brushy Basin Member of Morrison Formation
Jmw = Westwater Canyon Member of Morrison Formation
Jms = Salt Wash Member of Morrison Formation
Jb = Bluff Sandstone
Js = Summerville Formation
Jeu = Upper member of Entrada Sandstone
Jem = Middle member of Entrada Sandstone
Trw = Wingate Sandstone
A/S = analyst/source (SA = Anderholm, 1979; BB = Brod. 1979; DB = Brown, 1979; LF = Fleischhauer *in* Stone, 1979)
φ = phi
Slt = silt
VC = very coarse sand; VC-G = very coarse-granular
C = coarse sand
M = medium sand
F = fine sand
VF = very fine sand
S/C = silt and clay
Verb. = verbal description of mode
Agg. = aggregate
S_{ig} = inclusive graphic standard deviation of Folk (1974, p. 46)
S_g = graphic standard deviation of Folk (1974, p. 45); S_g is used where specific size of the 95th percentile is uncertain.
Verbal Sorting:
EP = extremely poor
P = poor
M = moderate
MW = moderately well
NA = not applicable
S/M = sand/mud ratio
* = samples contained appreciable amounts of aggregate; not plotted.

Part II Mineralogic Analyses

Anal. No.	Unit	A/S	Sample No.	Location (¼, sec, T,R)	Whole Rock (%) Frmwk	Cem	Mtx	Por	Framework (%) Q	F	RF	Size Range	Texture Mo	Srt	Rdns	Sph	Elong	TM	Class.
14	Kk (f)	SC	SJ-2-4-1	SW,21,29N,13W	82	6	9	3	80	17	3	VF-M	F	M	A-SR	P-G	VE-VEL	I	SA
15	Kk (f)	SC	SJ-2-4-2c	SW,21,29N,13W	77	6	7	10	64	28	8	VF-VC	M	P	A-R	P-G	VE-VEL	I	A
16	Kf	SC	SJR-59	(Navajo mine)	73	3	23	1	81	12	7	Cslt-F	VF	W	A-SR	P-E	VE-VEL	I	SA
17	Kf	SC	SJR-64	(San Juan mine)	63	31	4	2	67	26	7	Cslt-M	F	W	A-SR	P-E	VE-VEL	Sub-M	A/LA
18	Kpc	SC	SJ-1W-3-2	SW,32,30N,15W	69	7	19	5	74	25	1	Cslt-F	VF	VW	A-R	P-E	VE-VEL	I	A/SA
19	Kpc	SC	SJ-1W-3-6	SW,32,30N,15W	58	42	—	—	79	20	1	Cslt-F	VF	W	A-R	P-E	VE-VEL	Sup	SA
20	Kch	SC	SJR-68	SE,13,22N,13W	82	3	10	5	80	19	1	Cslt-M	M	W	SA-WR	P-G	VE-VEL	I	SA
21	Klv	SC	SJR-23	NW,16,19N,1W	83	6	3	8	83	13	4	Cslt-M	F	SA-R		P-E	VE-VEL	Sub	SA
22	Klv	SC	SJR-42	SE,19,19N,1W	76	21	2	1	84	13	3	Cslt-M	F	VW	A-R	P-G	VE-VEL	Sub	SA
23	Kmf	SC	SJR-4	NE,11,23N,1W	68	11	21	—	61	26	13	Cslt-F	VF	W	A-SA	P-G	VE-VEL	I	LA
24	Kmf	SC	SJR-5	SE,5,29N,16W	78	5	8	9	59	26	15	VF-C	M	M	SA-WR	P-E	VE-VEL	I	LA
25	Kmf	SC	SJ-1W-2B-34	SW,4,29N,16W	60	38	2	—	62	23	15	CSlt-M	F	M	A-WR	P-E	VE-VEL	Sub	LA
26	Kpl	SC	SJR-6	SE,5,29N,16W	76	15	4	5	70	16	14	VF-C	M	M	SA-R	F-G	SE	Sub	A
27	Kpl	SC	SJR-7	SE,5,29N,16W	78	12	4	6	67	20	13	VF-C	M	M	A-W	P-E	I	Sub	A
28	Kpl	SC	SJR-8	SE,5,29N,16W	62	38	—	—	60	26	14	VF-C	M	M	A-R	F-E	I-SE	Sub	A
29	Kpl	SC	SJR-9	SE,5,29N,16W	73	11	7	9	69	21	10	VF-C	M	MW	A-R	P-G	I	I	A
30	Kpl	SC	SJ-1W-1-2	NW,7,31N,16W	80	14	3	3	59	35	6	Slt-F	F	MW	A-SR	P-G	I	Sub	A
31	Kpl	BB	2B-16(1)	NW,15,13N,8W	61	5	32	2	57	29	12*	VF-M	F	W	SA	G-E	VE	I	LA
32	Kcda	BB	2B-7(2)	NW,14,14N,8W	67	16	13	4	64	31	5	Slt-C	F	W	SA	G-E	E	I	LA
33	Kmm	BB	2B-3(3)	NW,20,13N,8W	63	—	26	11	62	30	8	Slt-M	F	VW	SA	G	E	I	LA
34	Kg	SC	SJR-31	NW,9,30N,19W	80	8	3	9	74	14	12	Cslt-F	VF	VW	SA-R	F-E	VE-VEL	Sub	LA
35	Kg	SC	SJR-34	NE,26,28N,20W	86	8	3	4	70	28	2	F-C	M	W	SA-WR	P-E	VE-VEL	M	LA
36	Kg	SC	SJR-41	NW,21,30N,19W	87	4	2	7	73	23	4	F-C	C	W	SA-WR	P-E	VE-VEL	Sub	LA
37	Kg	BB	1B-3(4)	SW,16,14N,9W	84	3	7	6	56	41	2*	Slt-C	M	W	A-SA	G	E	M	LA
38	Kd	SC	SJR-2	SE,36,21N,1W	89	6	3	2	78	19	3	VF-M	F	VW	SA-WR	P-E	VE-VEL	Sub	SA
39	Kd	BB	1A-16(5)	SE,14,13N,10W	84	4	6	6	57	42	1	VF-M	F	W	A-SA	G-E	E-EL	M	A
Average framework (%) for Cretaceous sandstone aquifers analyzed									69	24	7								

Anal. no = analysis number
Unit abbreviations same as Part I
A/S = analyst/source (BB = Brod, 1979; DB = Brown, 1976; SC = Craigg, 1980)
Frmwk = framework
Cem = cement } parentheses denote
Mtx = matrix } composite
Por = porosity
Q = quartz pole
F = feldspar pole
RF = rock fragment pole (Folk 1974)
* = does not total 100%
Slt = silt
Cslt = coarse silt
VF = very fine sand
F = fine sand
M = medium sand
C = coarse sand
VC = very coarse sand
Grv = gravel
Mo = modal grain size
Srt = estimated sorting (Folk 1974)
VP = very poor
P = poor
M = moderate
MW = moderately well
W = well
VW = very well
Rdns = roundness (Krumbein and Sloss, 1956, figs. 4–9)
A = angular
SA = subangular
SR = subrounded
R = rounded
WR = well rounded
Sph = sphericity (Krumbein and Sloss, 1956, figs. 4–9)
P = poor
F = fair
G = good
E = excellent
Elong = elongation (Folk 1965)
VEL = very elongate
EL = elongate
SEL = subelongate
I = intermediate
SE = subequant
E = equant
VE = very equant
TM = textural maturity (Folk 1974)
I = immature
Sub = submature
M = mature
Sup = supermature
Class. = classification (Folk 1974)
A = arkose
LA = lithic arkose
SA = subarkose

Source: Stone and others (1983); portions of Tables 8 and 9.

trix, cement, and porosity. Continue by focusing on the make-up of framework grains alone, so that the sandstone may be properly classified by one of the prevailing schemes (for example, that of Folk 1974). If the material is too fine grained for this, use X-ray methods to identify the mineralogy.

The nature of various geologic units in the area may be readily compared if the texture and composition data are tabulated. For example, Stone and others (1983) compared the numerous sandstone aquifers in northwest New Mexico using such data (Table 2-4).

Completeness

Make sure that the geologic information you have compiled is complete. That is, make sure that the various sources of geologic information have been consulted, the major types of information have been gathered, and the appropriate tools have been used to evaluate and present the information. Furthermore, if any of the illustrations described here are used, they should be as complete as possible.

Maps should include the scale, a north arrow, a land grid (township and range, latitude and longitude, etc.), an explanation of any patterns or symbols used, and the contour interval, when appropriate. If the variation in some parameter over the study area is shown by contouring or subdivision, include the actual data so the map can be evaluated and possibly modified later, should additional information become available. When several maps of the study area are presented, it is very helpful to readers if they are at the same scale and drawn on the same base. Then readers' attention can be focused on the point of the illustration rather than on orienting themselves.

Various labels must be added to other types of illustrations to make them complete, as well. On cross sections it is useful to label the ends with compass directions (for example, "E" for east and "W" for west). When photographs are used in reports, indicate the direction the viewer is looking by a note in the caption (for example, view to west of . . . or looking west at . . .). In the case of geologic features (fault, dike, outcrop, etc.) it is also important to give the location (by land grid or distance and direction from a landmark along a known highway) in case the reader wants to visit the site.

REFERENCES

Asquith, G., and C. Gibson. 1982. *Basic well log analysis for geologists*. No. 216. Methods in Exploration Series. American Association of Petroleum Geologists.

Bredehoeft, J. D., and R. N. Farvolden. 1963. Disposition of aquifers in intermontane basins of northern Nevada. Extract of publication no. 64 of the IASH Commission of Subterranean Waters: 197–212.

Brod, R. C. 1979. Hydrogeology and water resources of the Ambrosia Lake–San Mateo area, McKinley and Valercia Counties, New Mexico: M. S. thesis, New Mexico Institute of Mining and Technology. 200 p.

Brown, D. R. 1978. Hydrogeology and water resources of the Aztec Quadrangle, New Mexico: M. S. thesis, New Mexico Institute of Mining and Technology. 174 p.

Brown, D. R., and W. J. Stone. 1979. Hydrogeology of Aztec Quadrangle, San Juan County, New Mexico. Hydrogeologic Sheet 1. New Mexico Bureau of Mines and Mineral Resources.

Craigg, S. E. 1980. Hydrogeology and water resources of the Chico Arroyo–Torreon Wash area, McKinley and Sandoval Counties, New Mexico: M. S. thesis, New Mexico Institute of Mining and Technology. 275 p.

Davis, S. N., and R. J. M. DeWeist. 1966. *Hydrogeology*. New York: John Wiley & Sons. 463 p.

Folk, R. L. 1974. *Petrology of sedimentary rocks*. Austin, TX: Hemphill's. 182 p.

Garner, H. F. 1974. The origin of landscapes—a synthesis of geomorphology. New York: Oxford University Press. 734 p.

Krumbein, W. C., and F. J. Pettijohn. 1938. *Manual of sedimentary petrography*. New York: Appleton-Century-Crofts. 549 p.

Krumbein, W. C., and L. L. Sloss. 1956. Stratigraphy and Sedimentation. San Francisco: W. H. Freeman and Company. 497 p.

Lance, J. O., Jr., G. R. Keller, and C. L. V. Aiken. 1982. A regional geophysical study of the western overthrust belt in southwestern New Mexico, west Texas and northern Chihuahua. *1982 Field Conference Guidebook,* p. 123–30. Rocky Mountain Association of Geologists.

Low, J. W. 1958. Subsurface maps and illustrations. In *Subsurface geology in petroleum exploration—a symposium,* p. 453–530. J. D. Haun and L. W. Leroy (eds.). Golden, CO: Colorado School of Mines.

O'Brien, K. M., and W. J. Stone. 1984. Role of geologic and geophysical data in modeling a Southwestern alluvial basin. *Ground Water* 22 (6):717–27.

Powell, J. W. 1895. Physiographic regions of the United States. Monograph 3: 6–100. National Geographic Society.

Powers, M. C. 1953. Comparison chart for visual estimation of roundness. *Journal of Sedimentary Petrology* 23:117–19.

Prothero, D. R. 1990. Interpreting the stratigraphic record. New York: W.H. Freeman. 410 p.

Rider, M. H. 1986. *The geologic interpretation of well logs*. London: Blackie/Halsted Press. 175 p.

Shimer, J. A. 1972. *Field guide to landforms in the United States.* New York: MacMillan. 272 p.

Stone, W. J. 1981. Hydrogeology of the Gallup Sandstone, San Juan Basin, northwest New Mexico. *Ground Water* 19:4–11.

Stone, W. J., F. P. Lyford, P. F. Frenzel, N. H. Mizell, and E. T. Padgett. 1983. Hydrogeology and water resources of San Juan Basin, New Mexico. Hydrologic report 6. New Mexico Bureau of Mines and Mineral Resources.

Thornbury, W. D. 1965. *Regional geomorphology of the United States*. New York: John Wiley & Sons. 609 p.

U.S. Geological Survey. 1969. *Landforms of the United States*. Booklet. U.S. Geological Survey.

Walton, W. C. 1970. *Ground water resource evaluation*. New York: McGraw-Hill. 664 p.

CHAPTER 3

Characterizing the Geologic Setting

Once you have compiled the available information, the next step is to evaluate it and use it to describe the geologic phenomena in the study area. This mainly involves characterizing the stratigraphic framework and structural conditions. However, depending on its potential hydrologic significance, you may wish to include a description of the local geomorphic setting as well.

GEOLOGIC PHENOMENA

The various topics included in the geologic characterization may be categorized as either conditions or features (Table 3-1). Geologic conditions include such topics as the stratigraphic sequence and general stability or activity of the area. Geologic features include the products of various structural and geomorphic processes.

The rock record or stratigraphic sequence is the medium within which the hydrologic system operates. Thus, it is important to determine as much about it as possible, as early as possible. Several basic questions should be answered under this topic.

TABLE 3-1 Geologic Phenomena That May Be Characterized in Hydrogeologic Studies

Geologic Conditions	Geologic Features
Stratigraphic sequence	Unconformities
Tectonic stability	Structures
Subsidence	Folds
Uplift	Faults; Joints
Seismic hazard	Depositional landforms
Volcanic activity	Sedimentary
Heatflow	Volcanic
Erosional/depositional	Erosional landforms
Prone to flooding	Fluvial
	Glacial
	Karst
	Eolian
	Marine

What geologic materials underlie the study area?
What is their thickness and extent?
What are their lithologic characteristics?
How uniform are these characteristics across the area?
Are these materials in contact with something different in adjacent areas?
What is the nature of the contacts between units (unconformable, intertonguing, faulted, etc.)?

Answers to these questions may be changed or supplemented as the study progresses.

The geologic stability of the area is also an important part of the general setting. You should answer several questions about this topic.

Is the area tectonically (seismically) active?
Is it subject to volcanism?
Is it an area of high heat flow?
Is it being actively incised or lowered by erosion?
Is it being actively buried by deposition?
Is it prone to flooding (riverine or pluvial)?

Although less significant, perhaps, for hydrogeology than the rock record or stratigraphic sequence, such conditions are critical in siting or monitoring waste-disposal facilities. Although heat flow is mainly of interest in hydrogeologic studies of geothermal resources or areas, it has been used to investigate ground-water movement (for example, Wade and Reiter 1994; Reiter and Jordan 1996).

Structural features are those resulting from deformation of the crust. Familiar examples are folds, faults, and joints. Of particular interest are the type, extent, and orientation of such features. In the case of faults, the materials juxtaposed by the displacement is also important. A knowledge of these characteristics is essential when determining geologic controls of hydrologic phenomena.

Geomorphic features are those resulting from petrogenic, erosional, or depositional processes at or near the earth's surface. Petrogenic features are those resulting from rock-forming processes. Igneous activity produces a variety of intrusive and extrusive features. Although extrusive features are formed at the surface and thus are more prominent, intrusive features may be exhumed by erosion and also become a part of the landscape. Sedimentary and diagenetic processes may result in such features as reefs, salt domes, or sandstone dikes. All of these features may exert control on hydrologic conditions. Erosional and depositional features are those produced by the work of wind, water, ice, and gravity. As with the features previously mentioned, location, size, and orientation are significant. The geologic units affected by these features (if erosional) or associated with them (if depositional) are critical to our understanding of regional or areal hydrology.

STRATIGRAPHIC FRAMEWORK

A major part of the geologic setting of an area is its stratigraphy, that is, the geologic units present, their make up, and their relationship to each other. In a report, a section on the geologic setting generally starts with a discussion of the sequence, followed by a description of units by name.

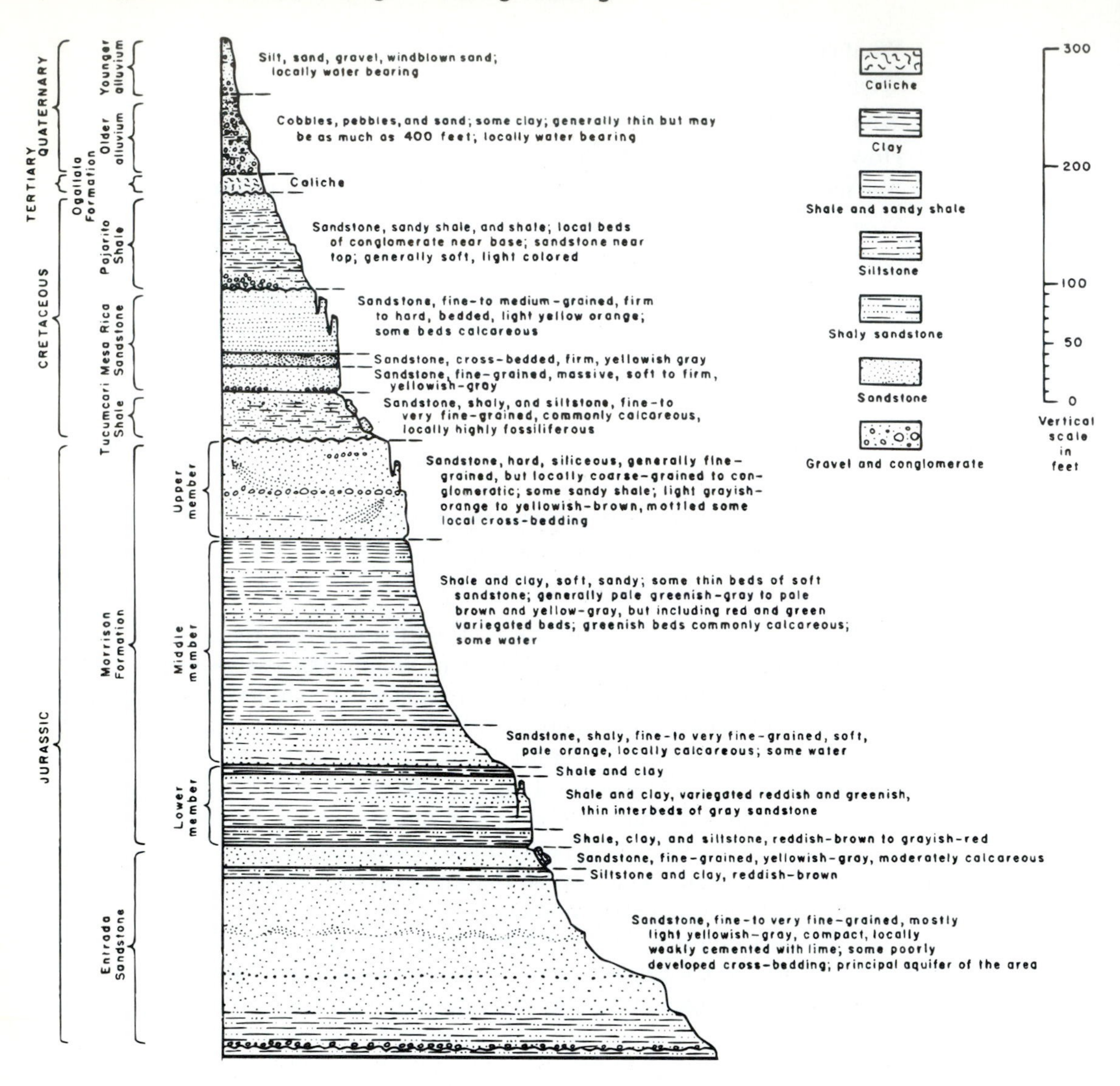

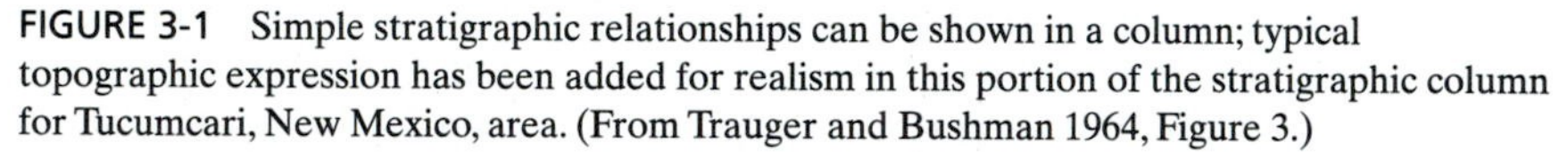
FIGURE 3-1 Simple stratigraphic relationships can be shown in a column; typical topographic expression has been added for realism in this portion of the stratigraphic column for Tucumcari, New Mexico, area. (From Trauger and Bushman 1964, Figure 3.)

Stratigraphic Sequence

It is essential to first present the rock record for the area. This is most easily done with a geologic column (Figure 3-1). The column may be actual, if the site is small or if there is little variation in stratigraphy across it, or idealized, if different units occur in different parts of the study area. If these differences are great and relationships complex, a cross section constructed along a line selected to show the maximum variation present may be required instead of a column (Figure 3-2). Additionally, some of the subsurface maps described in Chapter 2 may be warranted.

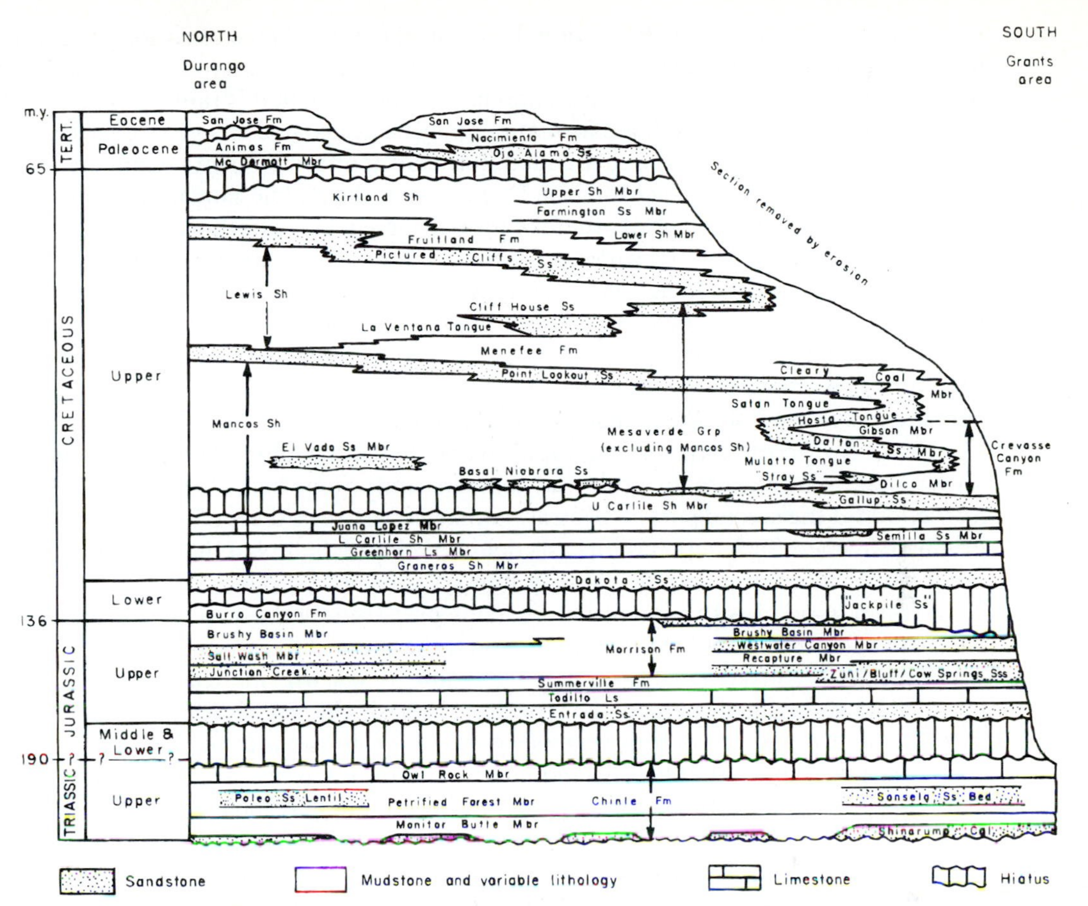

FIGURE 3-2 Complex stratigraphic relationships cannot be shown in a simple column; this cross section shows the intertonguing of units in the San Juan Basin, New Mexico. (From Stone and others 1983, Figure 6.)

Stratigraphic Nomenclature

Naming stratigraphic units is required for geologic bookkeeping and clear communication. Stratigraphic nomenclature is as strictly regulated as biological taxa. There is an International Stratigraphic Code (Hedburg 1976), but many countries have their own systems. The governing rules for geologic names in the United States are set by the North American Commission on Stratigraphic Nomenclature (1983). Hydrogeologists working in North America are encouraged to examine this code and become familiar with it. The code is reproduced in some stratigraphy textbooks (for example, Prothero 1990). Useful suggestions on using stratigraphic nomenclature were given by Owen (1987).

Although it is not practical to duplicate the code here, some general summary remarks are appropriate. Formal stratigraphic names consist of two parts: Morrison Formation, Dakota Sandstone, etc. The first part is a unique geographic name, usually derived from the locality where the unit is best exposed or was first described. The second part is a term to indicate stratigraphic rank. The most fundamental rank and basic geologic mapping unit is the formation. It is a distinct assemblage of rocks or deposits having a common lithologic character or origin. The major rock type may be used in place of the word *formation* for the second part of the name in this rank: Dakota Sandstone. Two or more closely related formations may be designated as a group: Mesaverde Group. Formations may also be subdivided (formally or informally) into members: Westwater Canyon Member of the Morrison Formation or the middle sandstone member of the Moreno Hill Formation. Note that the formal name is capitalized while the informal name is not. Distinct material within a formation or member may be formally or informally designated as a bed, such as the Huerfano Marker Bed of the Lewis Shale (a thin bentonite layer, easily recognized on electric logs). Although a unit must be mappable at a scale of common topographic maps in order to be designated as a formation, that prerequisite is not necessary for ranks smaller than formation. Nonetheless, some members are of sufficient thickness that they are mapped.

Take care to use correct stratigraphic names in hydrogeologic reports. You should always determine and use the preferred name for units. Some hydrologists use the term *formation* for the second part of all stratigraphic names, regardless of how the unit was originally named. Such imprecision casts doubt on the credibility of the entire report. In other words, if the stratigraphy is sloppy, how good is the hydrology?

It is a good idea to consult the lexicons prepared by the USGS (for example, Keroher and others 1966) to learn more about the units in the study area. These lexicons give the preferred name for the unit, its rank, its age, who named it, the reference in which it was originally named, the type area, its basic characteristics, and subsequent modifications by other workers, if any (Figure 3-3). It is useful to photocopy the lexicon entry for each unit in the area and make a file for reference during the study. Because these lexicons were prepared some time ago, there may have been changes that they do not include. For example, the name of the unit shown in Figure 3-3 has recently been changed to the Puye Formation. It is a good idea to consult the State Geological Survey for the current usage. If possible, the type area should be visited and the sequence examined first-hand, referring to measured sections from the original reference, so that the major characteristics of the units become clear and familiar. If some holes are to be drilled in an unfamiliar area, it is instructive to examine cuttings and logs from nearby wells that penetrated the same sequence. These materials may be available from federal, state, county, or municipal agencies, but most often are archived at the State Geological Survey.

Stratigraphic Unit Description

In addition to giving the stratigraphic column or a cross section of the area, a hydrogeologic report should contain a thorough description of each unit present or at least those relevant to the study. Although geologists normally think in terms of oldest to youngest (bottom to top), hydrologic reports often present units in the order that they would be encountered in drilling: youngest (shallowest) to oldest (deepest). In any case, the order employed should be identified at the outset.

Adopting a standard set of parameters to be described and sticking to it will not only ensure adequate and consistent coverage, but will be greatly appreciated by users of the information. Several items are suggested for inclusion in unit descriptions:

LEXICON OF GEOLOGIC NAMES OF UNITED STATES 3165

Puye Conglomerate (in Santa Fe Group)

Puye Gravel (in Santa Fe Group)

Pliocene (?) : North-central New Mexico.

H. T. U. Smith, 1937, (abs.) Geol. Soc. America Proc. 1936, p. 103; 1938. Jour. Geology, v. 46, no. 7, p. 937 (fig. 4), 950. Quaternary formations in Abiquiu quadrangle comprise Canjillion till, Canones andesite, Vallecito basalt (all new), and Black Mesa basalt, which were poured out on high-level erosion surfaces, and Puye gravel (new), which overlies an erosion surface of intermediate level and is overlain, locally, by Bandelier rhyolite and Santa Clara basalt (both new).

E. H. Baltz and others, 1952, New Mexico Geol. Soc. Guidebook 3rd Field Conf., p. 12 Ancha formation (new) is eastward extension of Puye gravel.

Brewster Baldwin, 1956, New Mexico Geol. Soc. Guidebook 7th Field Conf., p. 118 (fig. 2), 119. Mentioned in discussion of upper unit of Santa Fe group. Puye gravel, Ancha formation, and Tuerto gravel all rest with angular unconformity on deformed beds of Tesuque formation. These units of gravel are 500, 300, and 150 feet in maximum thickness, respectively. In Buckman area, Ancha formation appears to intertongue with Puye gravel. Early Pleistocene.

Mapped in valley of Chama River, Rio Arriba County.

FIGURE 3-3 Typical USGS lexicon entry; note the types of information routinely provided. (From Keroher and others 1966, p. 3,165.)

Preferred name
Age
Original reference
Type area
Subdivisions, if any
Lithologic parameters
- Rock or deposit type
- Texture (grain size, shape, sorting)
- Mineralogy
- Bedding and primary structures
- Other characteristics (cement, concretions, etc.)

Thickness
Extent
Nature of contact with underlying unit

It is also helpful to provide a general statement on where the examined unit appears in the sequence and its relationship to major geologic features. For example, "The Cliff-house Sandstone is the uppermost unit in the Mesaverde Group and forms the eastern flank of the Hogback Monocline." (Stone and others 1983).

The first five items in the list come from the USGS lexicon. The remaining lithologic and stratigraphic characteristics should be drawn from field work and laboratory data or, if the opportunity for field and lab work is limited, from information in the geologic literature on the area.

STRUCTURAL CONDITIONS

Once the geologic sequence is described, all that remains to do is to characterize any deformation that may have altered it. This includes a discussion of major folds, faults, and fractures. Structure maps and cross sections are an excellent means of illustrating folding and/or faulting in the area and should be employed if they are available in the literature or if subsurface data permit their construction. The following brief discussion of features to include assumes some training in structural geology; readers lacking this should consult a standard structural geology text (for example, Billings 1954).

Major Features

First, you should describe the location of the study area relative to a major structural feature, if there is one. For example, the study area might lie at the edge of a graben, atop a dome, or within the region of a major thrust fault. If the feature has a formal name (e.g., Rio Grande rift, Black Hills uplift, Roberts Mountains thrust), it should be mentioned. The feature, or part of it associated with the study area, should be described.

Folds

Folds are wrinkles in the stratigraphic carpet. Such features may be significant hydrologically. In describing them, identify the basic kind of folds present (anticline, syncline, monocline), determine the specific types based on dips of opposing limbs or the relationship of their axial plane to vertical (symmetrical, asymmetrical, overturned, recumbent), and give their orientation (compass direction). Any plunge (single, double) and its direction(s) should also be described. Thus, a complete fold description might read, "north plunging, asymmetrical syncline" or northwest-trending symmetrical dome."

Faults

Faults are breaks in rocks along which there has been movement. If there is faulting in the study area, it may be important hydrologically. Thus, the type (dip fault, if movement is parallel to dip; strike fault, if movement is parallel to strike; oblique fault, if movement is not parallel to either), orientation (compass direction of trend), and nature of movement (normal, reverse, thrust) should be described. If a fault zone is exposed in outcrop, a description of its characteristics is most helpful. For example, is it narrow or wide? Is it open or filled with gouge? Does it juxtapose units of similar or markedly different hydrologic properties?

Joints

Joints are smooth fractures or partings that break or interrupt the continuity of rocks, but along which there has not been appreciable movement. Because these features can provide the main source of porosity and important pathways for water movement in well-consolidated or crystalline rocks, note carefully their presence, character, and orientation.

Major characteristics of joints include width, length, and density. The width of joints that have not been modified by erosion (dissolution) or deposition is generally measured in inches. The length of joints may range from a few feet to thousands of feet. Because joints always occur in groups, their density is also a useful parameter to describe. The interval between them can range from inches to hundreds of feet.

The orientation of joints can be vertical, horizontal, or anything in between. They may be classified by their attitude relative to bedding: strike joints (parallel to strike of beds or other fabric in rock), dip joints (parallel to dip of beds or other fabric), oblique or diagonal joints (attitude falls between that of strike or dip of host rock or host-rock fabric), and bedding joints (parallel to beds).

If a number of joints are parallel, they are said to form a set. Two or more joint sets are called a joint system. If these are recognized in the study area, describe them, especially their number and orientation. Where joints are abundant, a good way to depict their importance in an area is to map them (Figure 3-4).

GEOMORPHIC SETTING

Nonstructural features may be prominent in the study area, as well. They include petrogenic as well as erosional and depositional features. Together with the stratigraphy and structure, these geomorphic features complete the geologic setting. They should be described not only for completeness, but because they, too, may influence area hydrology.

Petrogenic Features

Landscape features resulting from rock-forming processes are mainly of intrusive or extrusive igneous origin. Intrusive features include batholiths, stocks, dikes, and sills (Figure 3-5). Those features of extrusive origin are caldera, volcanic cones, lava flows, and ash-fall blankets. Note and describe the presence of any such features in the study area. Of special interest are their size, shape, orientation, and relation to stratigraphy and structure. Of what kind of rock are they composed? What stratigraphic or map units do they represent and cross-cut?

Erosional and Depositional Features

Many geomorphic features are erosional or depositional in origin. Erosional landforms include such features as valleys, terraces, canyons, mesas, buttes, horns, cirques, and sinkholes. Examples of depositional features are alluvial fans, dune fields, moraines, eskers,

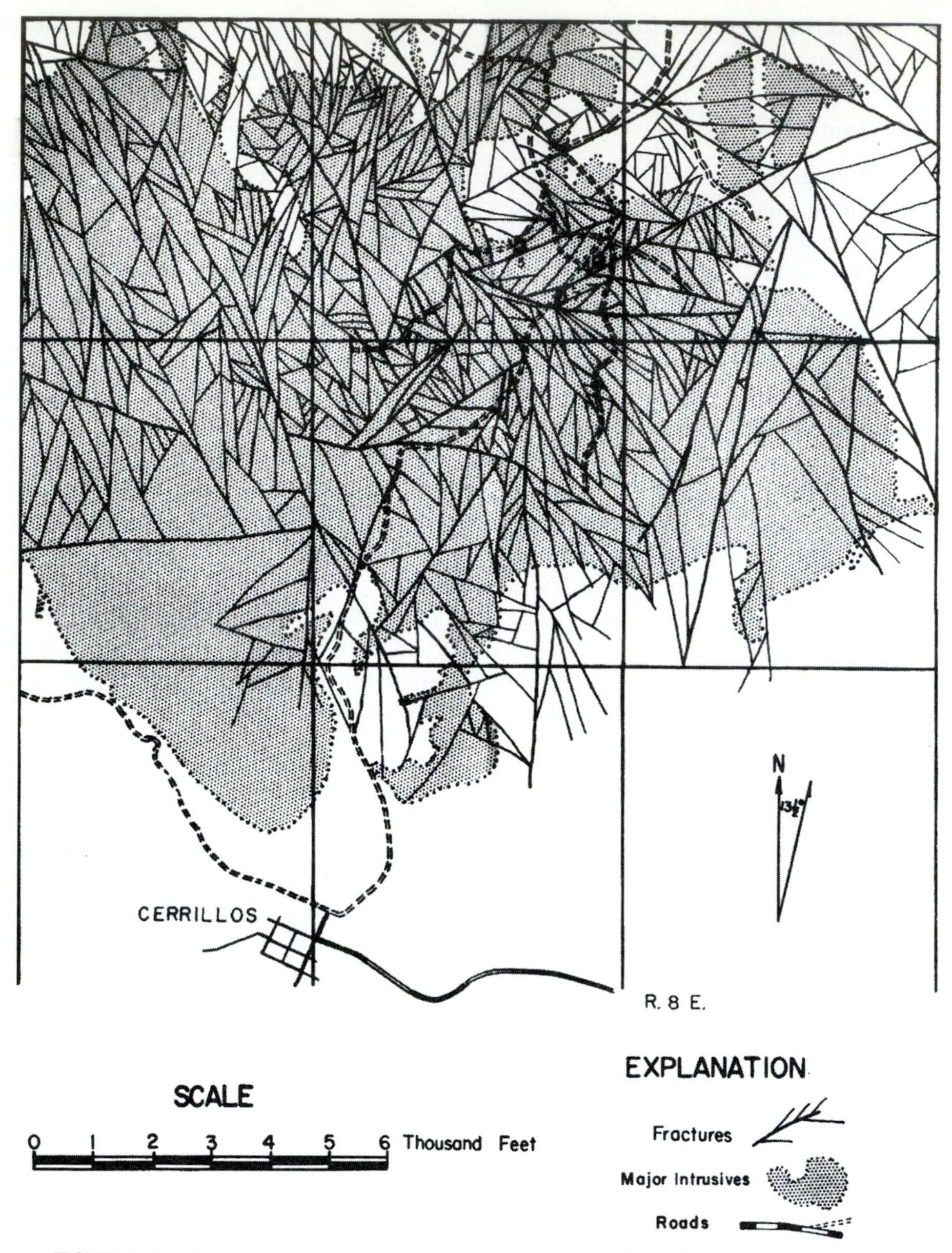

FIGURE 3-4 Map of joints in the Cerrillos area, New Mexico. (From Disbrow and Stoll 1957, portion of Plate 2.)

FIGURE 3-5 Dikes are hydrologically significant petrogenic features; one of several dikes in the northeastern part of the San Juan Basin, northwestern New Mexico. (From Stone and others 1983, Figure 85.)

and deltas. Describe such features, emphasizing location, dimensions, extent, and orientation, if any. The geologic units affected (if erosional) or associated (if depositional) is also important.

The section on geologic setting is one of the most important parts of a hydrogeologic report. Make it as complete and correct as possible. As new information becomes available from drilling or field work, the report should be updated.

REFERENCES

Billings, M. P. 1954. *Structural geology.* Englewood Cliffs, NJ: Prentice Hall. 514 p.

Disbrow, A. E., and W. C. Stoll. 1957. Geology of the Cerrillos area, Santa Fe County, New Mexico. Bulletin 48. New Mexico Bureau of Mines and Mineral Resources. 73 p.

Hedburg, H. D., ed. 1976. *International stratigraphic guide—a guide to stratigraphic classification terminology and procedure.* New York: John Wiley & Sons. 200 p.

Keroher, G. C., and others. 1966. Lexicon of geologic names of the United States for 1936–1960. Bulletin 1200, 3 parts. U.S. Geological Survey. 4,341 p.

North American Commission of Stratigraphic Nomenclature. 1983. *North American stratigraphic code.* 67(5): 841–75. American Association of Petroleum Geologists.

Owen, D. E. 1987. Commentary—usage of stratigraphic terminology in papers, illustrations and talks. *Journal of Sedimentary Petrology* 57(2): 363–72.

Prothero, D. R. 1990. *Interpreting the stratigraphic record.* New York: W.H. Freeman. 410 p.

Reiter, M., and D. L. Jordan. 1996. Hydrogeothermal studies across the Pecos River Valley, southeastern New Mexico. *Bulletin* 108(6): 474–756. Geological Society of America.

Stone, W. J., F. P. Lyford, P. F. Frenzel, N. H. Mizell, and E. T. Padgett. 1983. Hydrogeology and water resources of San Juan Basin, New Mexico. Hydrologic report 6, New Mexico Bureau of Mines and Mineral Resources. 70 p.

Trauger, F. D., and F. X. Bushman, 1964. Geology and ground water in the vicinity of Tucumcari, New Mexico. Technical report 30. New Mexico State Engineer. 178 p.

Wade, S. C., and M. Reiter. 1994. Hydrothermal estimation of vertical ground-water flow, Canutillo, Texas. *Ground Water* 32(5): 735–42.

CHAPTER 4

Geologic Materials as Aquifers

The term *aquifer* is used to refer to any material that both stores and transmits water and whose saturated portion yields useful quantities of water. Obviously, "useful" includes suitable water quality, as well. By contrast, the terms *aquitard* and *aquiclude* refer to materials that store but do not readily transmit or yield useful quantities of water. *Aquifuge* denotes a material that neither stores nor transmits water.

THEORY VERSUS PRACTICE

Theoretically, all porous materials below the water table are saturated or water bearing, regardless of their ability to yield the water they hold. Ground water includes all water in the saturated zone. Except for local perched conditions, the saturated zone may be viewed as a continuum, persisting across basins, through mountains, and under plateaus, to some depth below which appreciable porosity does not exist, due to lithostatic pressure.

In practice, however, ground-water occurrence is often described in terms of aquifers. This practice is merely an attempt to direct attention toward the most favorable sources of ground water in an area. However, do not conclude that ground water occurs only in aquifers or that only the aquifers contain water. The continuum concept should dispel such notions.

The use of the term *aquifer* is obviously subjective. Both unconsolidated and consolidated materials vary in their ability to store and transmit water. What is a useful yield for one purpose may not be for another (e.g., domestic versus municipal supplies). Thus, what is considered an aquifer in one area might not rate a second look in another because of the prevailing water use. For example, a saturated "dirty" (silty or clayey) sandstone might be a welcome target for a windmill in the Australian Outback, but be an utter disappointment for an irrigation operation in the American Midwest.

Some have argued that since ground water occurs throughout the saturated zone and flows across stratigraphic and even structural boundaries, the aquifer concept is invalid, or at least misleading. It need not be if you remember that the designation of units as aquifers is merely a way of identifying productive materials, not a way of explaining ground-water hydrology.

Consider the situation shown in Figure 4-1. In this actual case, a stock well that tapped north-dipping Tertiary volcanics went dry. When productive, the well had yielded only 3 gallons per minute (gpm). However, that yield was sufficient for a windmill. Foolishly, the rancher drilled a new well next to the old, dry one and predictably got nothing. It was suggested that he have the rig move south and drill in the Cretaceous coal measures lying beneath the volcanics, targeting the several marine sandstones in that in-

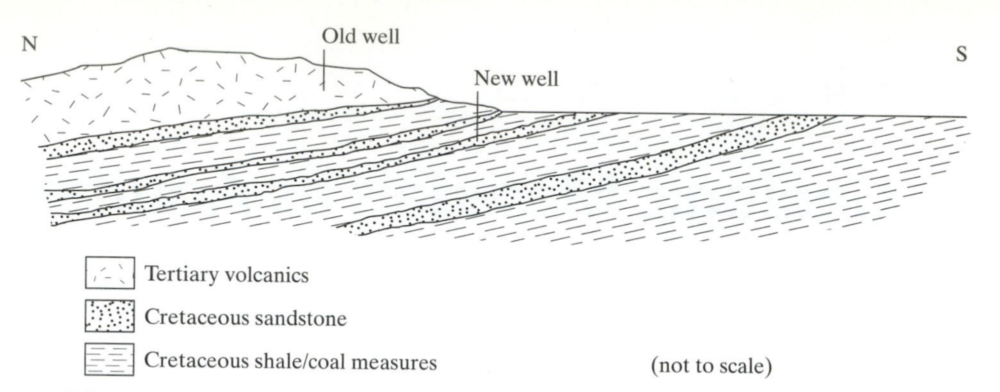

FIGURE 4-1 Schematic cross section illustrating the significance of the term "aquifer"; the well in volcanic materials had a yield of 3 gpm, but the well in marine sandstones had a yield of 30 gpm.

terval. This was done and a more than adequate yield of 30 gpm was obtained. Why did the new well yield 10 times that of the old well? The marine sandstone was a better aquifer than the volcanic rock.

The meaning of *aquifer* becomes even clearer if you visualize other scenarios where one well yields X gpm and an adjacent one yields 10X gpm. Some possibilities are shown in Figure 4-2. You may be able to think of other examples as well.

In some hydrogeologic reports, the writer seems to confuse the concepts of aquifer and water table. In such reports, the water table is labeled on figures as the "top of the aquifer." This thinking has led to statements such as, "seasonally the aquifer advances down or retreats up a canyon." *Danger! Moving aquifers!* Clearly, water level may rise and fall, but the aquifer doesn't move.

In preferred usage, *aquifer* applies to the entire water-bearing geologic unit (or entire group of water-bearing units), not just its saturated portion. Since aquifers are essentially reservoirs, we may compare them to partially filled water tanks. The water level at some position within it is obviously not the top of the tank. Such an analogy should make it easy to remember that: (1) the aquifer is the entire available reservoir, and (2) the water table is not the top of the aquifer.

HYDROSTRATIGRAPHIC UNITS

Geologic materials in the stratigraphic column may be classified according to their hydrologic behavior or function (Figure 4-3). Some hydrogeologic units may coincide exactly with rock-stratigraphic units. Others may involve more than one rock unit and require grouping; that is, some formations may act together as a single aquifer. Still other hydrogeologic units may involve only parts of formal rock units and require splitting, that is, some members of a formation may constitute aquifers and other members may behave as aquitards (Figure 4-4, p. 48).

The name of a hydrogeologic unit is generally derived in one of three ways: by its stratigraphy, depth, or lithology. In some cases, the formal stratigraphic unit name is applied, such as Ogallala aquifer. If there is no appropriate formal name, and there are

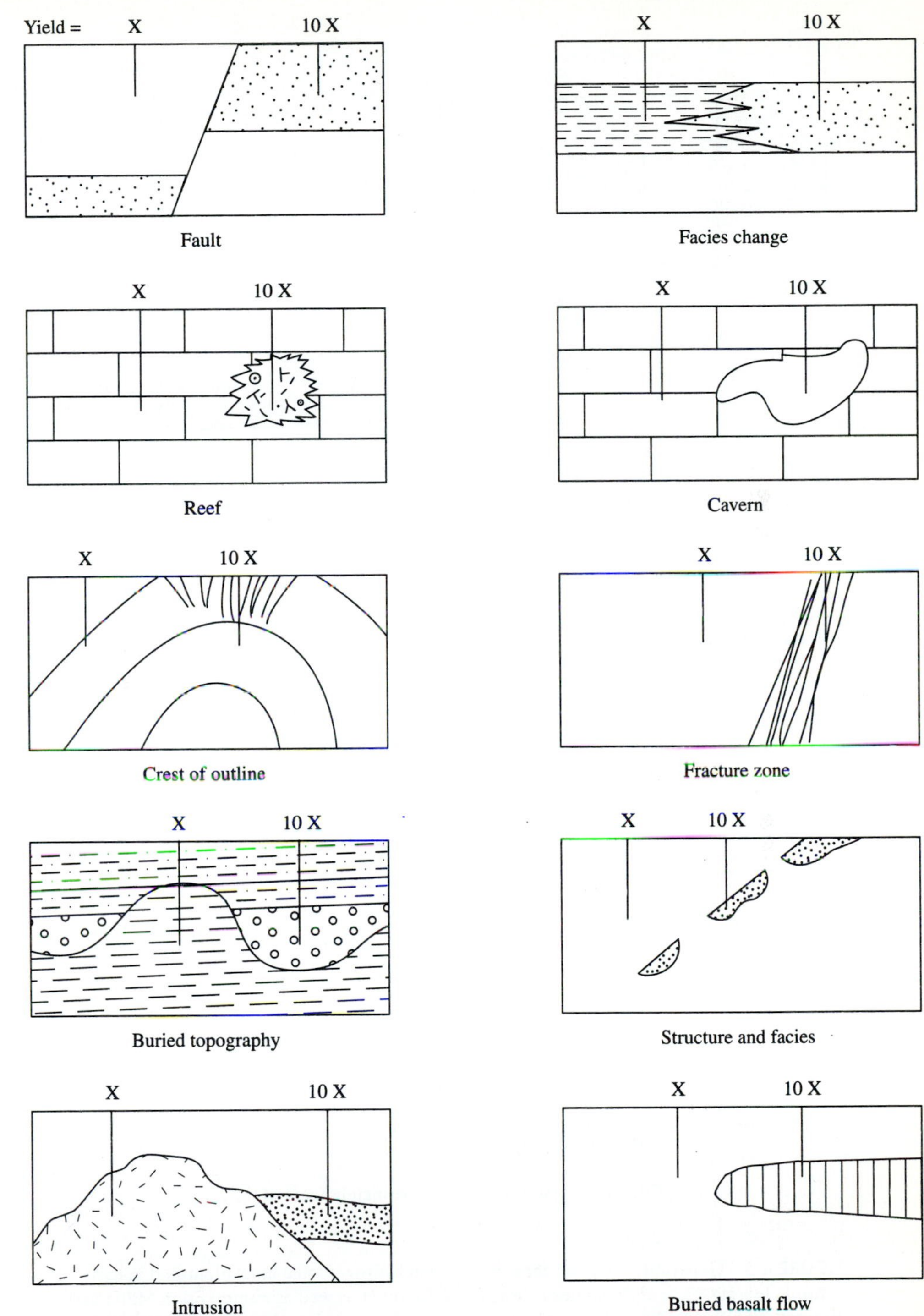

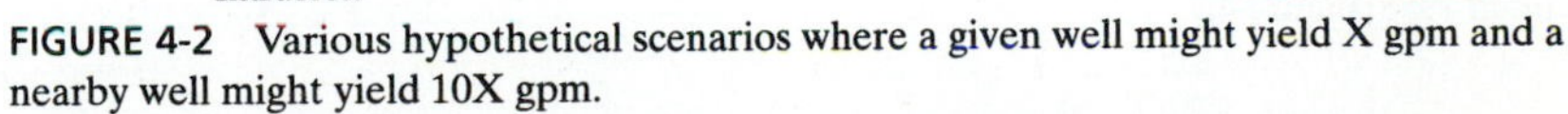
FIGURE 4-2 Various hypothetical scenarios where a given well might yield X gpm and a nearby well might yield 10X gpm.

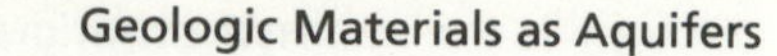

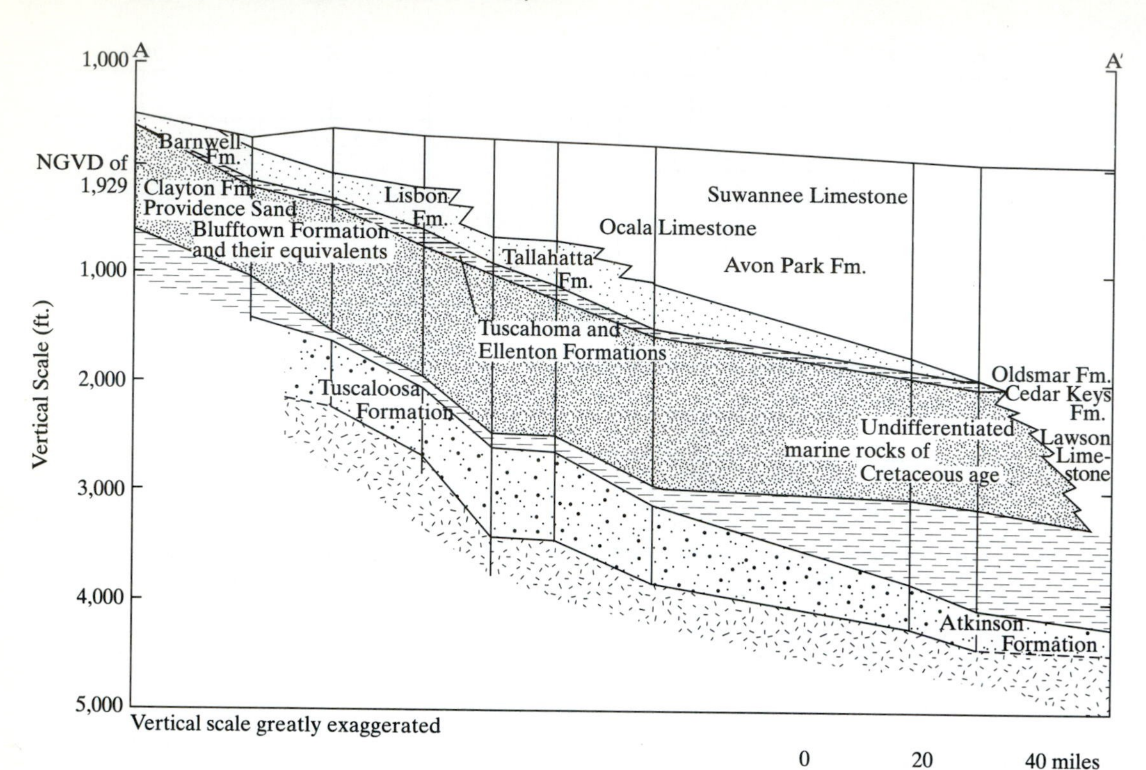

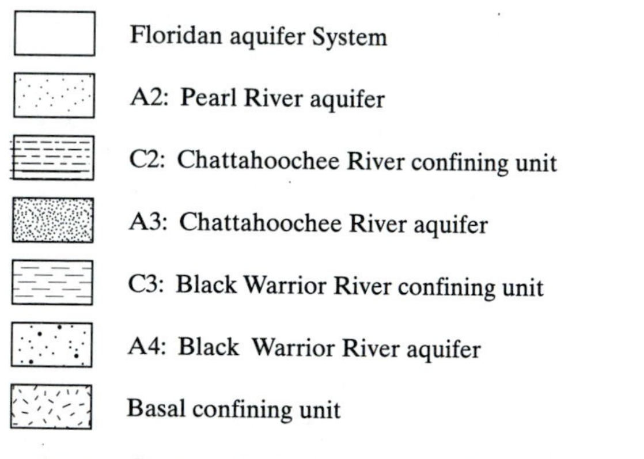

FIGURE 4-3 Generalized section showing regional hydrogeologic units identified in east-central Georgia. Note alphanumeric designation for units, as well as name. (From Miller and Renken 1988, Figure 4.)

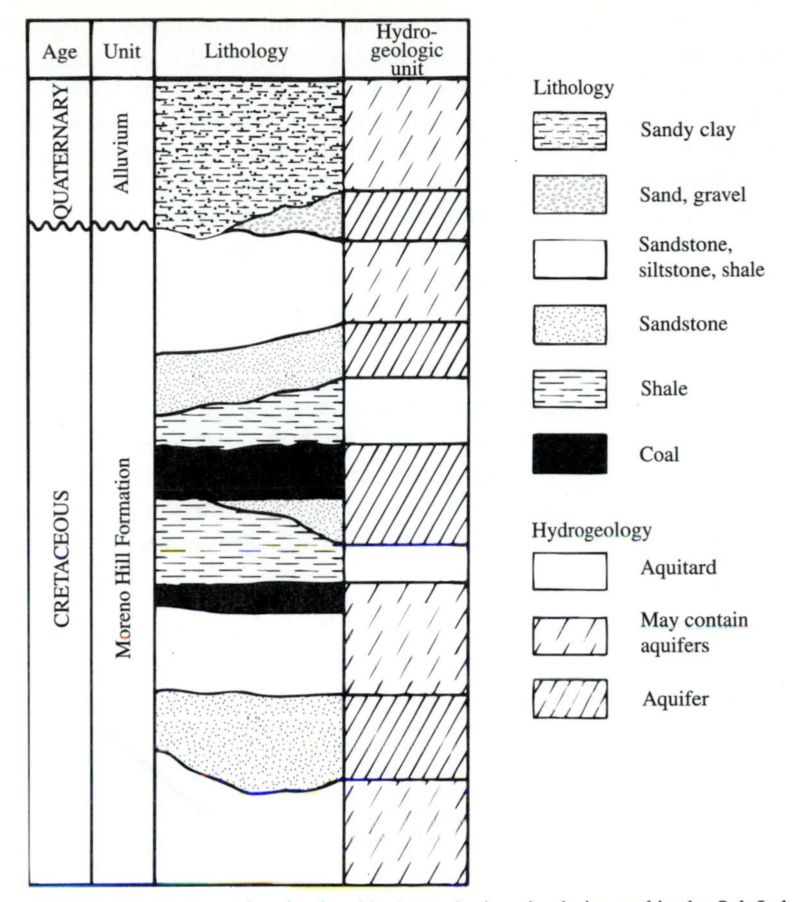

FIGURE 4-4 Column showing local hydrogeologic units designated in the Salt Lake coal field, west central New Mexico. (From Stone and McGurk 1987, Figure 5.)

multiple hydrostratigraphic units, depth might be used: shallow aquifer, intermediate confining bed, etc. Such names might be modified by incorporating the scale and hydrologic condition: shallow perched aquifer, regional unconfined aquifer, middle artesian aquifer, etc. Also, a name may be derived from the nature of the geologic material involved: the alluvium aquifer, the outwash aquifer, carbonate aquifer, basalt aquifer, etc. Note that these names are nouns, as are the formal stratigraphic names.

Some workers have, probably unwittingly, applied a fourth, less desirable approach to naming aquifers. For example, they use "alluvial aquifer," rather than "alluvium aquifer." In this case, the name is an adjective suggesting genesis. This genetic approach is not favored, but if employed for one aquifer, it should be employed for all the aquifers in the area, for parallel usage. Applying this scheme to the previous examples, the outwash would be named the "glacial aquifer," the carbonate rocks would become the "marine aquifer," and the basalt would be called the "volcanic aquifer." Whatever approach to naming aquifers is used, it is a good idea to be consistent.

In most of these cases, hydrogeologic units are treated informally. That is, there is no precise definition giving a specified criterion for the top and bottom, or a prescribed yield, so that it may be distinguished from other intervals. With a little common sense, such informal names should not present a problem. However, it has long been recognized that assignment and use of formal names can cause confusion, unless some standard guidelines are established and followed. What is the legal status of formal hydrostratigraphic units? A special committee of the Hydrogeology Division of the Geological Society of America was formed in the early 1980s to investigate the classification of water-bearing deposits within the North American Code of Stratigraphic Nomenclature (Seaber 1986 and 1988). Also, as might be expected, the USGS has been interested in this problem for some time. A comprehensive source of information and guidelines on aquifer nomenclature is the USGS Open-file Report by Laney and Davidson (1986). The USGS has developed formal nomenclature for hydrogeologic units in various regions. For example, Miller and Renken (1988) have divided the clastic sedimentary units of the Coastal Plain of the southeastern United States into four regional aquifers separated by three confining units (see Figure 4-3). Currently, however, the stratigraphic code has not yet been revised to include hydrostratigraphic units.

AQUIFER MATERIALS

Various types of materials can be aquifers. The only requirement is that materials are porous, so they can store water, and that the pores are connected, so they can transmit it. Aquifers may be classified by porosity type, lithology, and genesis.

Types by Porosity

Aquifer materials may be divided into two groups on the basis of porosity type: those having primary porosity and those having secondary porosity. Primary porosity is open space formed along with the geologic material. Secondary porosity is that formed after the geologic material is in place. As recognized by Meinzer (1923), the amount of porosity can vary within these broad categories, and some geologic media have both (Figure 4-5).

Primary porosity is most common in granular or clastic sediments and sedimentary rocks. Because the open space is between grains, *intergranular porosity* is synonymous with primary porosity. Aquifers of this type include gravel, sand, silt, and their lithified counterparts (conglomerate, breccia, sandstone, and siltstone).

Primary porosity may also occur in crystalline or nonclastic sediments. Carbonate sediments are poorly compacted and have high primary porosity: 40%–70% (Bathurst 1975). Such material may constitute the aquifers in oceanic islands.

Unlike primary porosity, secondary porosity is essentially limited to consolidated materials. It is most common in brittle rocks. For example, dense crystalline rock (igneous, sedimentary, or metamorphic), lacking any primary porosity, may nonetheless be an aquifer because of secondary porosity, such as fracturing and jointing. *Fracture porosity* is synonymous with secondary porosity.

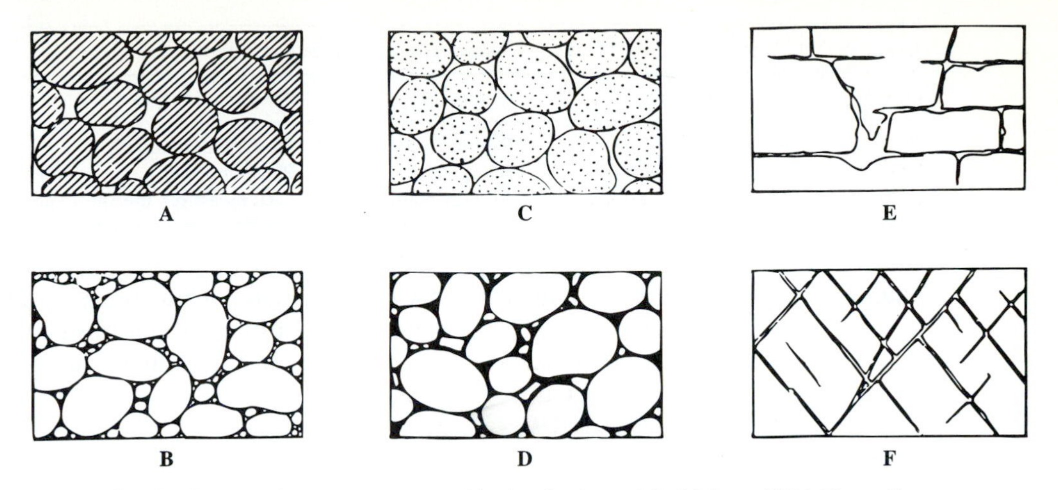

FIGURE 4-5 Porosity types as presented in the classic work by Meinzer (1923, Figure 1): (a) well-sorted sedimentary material with high intergranular porosity; (b) poorly sorted sedimentary material having low intergranular porosity; (c) well-sorted sedimentary material in which the clasts are also porous for a very high total porosity; (d) well-sorted sedimentary material in which porosity has been diminished by precipitation of mineral matter between grains; (e) rock made porous by dissolution; (f) rock made porous by fracturing.

Secondary porosity can enhance the hydrologic properties of terrigenous clastic rocks. For example, although shale has a considerable capacity to store water, due to primary intergranular porosity, the pores are so small that they provide resistance to water movement. Also, cementation may greatly reduce, if not eliminate, the primary porosity of even coarse clastic sedimentary materials. However, fractures can provide secondary porosity in both fine-grained and plugged materials, enhancing their aquifer potential.

Secondary porosity is especially important in carbonate aquifers. Cementation of carbonate sediments by calcite during diagenesis reduces porosity to <5% for most carbonate rocks (Bathurst 1975). Fracturing can greatly improve both porosity and permeability. These properties may be even more enhanced by dissolution of the rock material. Thus, the main porosity of carbonate or evaporite strata may be provided by enlargement of fractures or bedding planes through dissolution.

Types by Lithology

Aquifers may also be classified by sediment or rock type. Under such a scheme, potential aquifer materials include gravel, sand, silt, conglomerate, breccia, sandstone, siltstone, shale, coal, limestone, basalt, pumice, granite, quartzite, slate, gneiss, schist, etc. This is a refinement of the classification based on porosity. However, instead of lumping geologic materials into broad groups of primary or secondary porosity, the differences within these categories are distinguished. For example, although sand and silt (or sandstone and siltstone) both generally have primary or intergranular porosity, the coarser grain size of the sand or sandstone usually makes it superior to silt or siltstone as an

TABLE 4-1 Comparison of Some General Characteristics of Sandstone Aquifers

Origin	Texture	Geometry	Associated Strata
Fluvial/glacial	Variable	Ribbon, wedge	Floodplain muds/till
Eolian	Very uniform	Blanket	Playa deposits
Marine	Fairly uniform	Ribbon, blanket	Coal measures, marine shale, limestone

aquifer material. Similarly, crystalline rocks whose secondary or fracture porosity is (or may be) enhanced by dissolution (e.g., limestone or gypsum) are more significant as aquifers than those not likely to be modified in this way (e.g., granite or quartzite).

Types by Genesis

Many hydrogeology textbooks approach aquifers from an origin point of view. Some texts organize their coverage of aquifers in terms of broad genetic types: sedimentary, igneous, and metamorphic. Others, recognizing that sedimentary materials make the best aquifers, approach this group in more detail, still using origin: fluvial, glacial, eolian, and marine. Nonsedimentary rocks may also be distinguished by genesis: volcanic, plutonic, and metamorphic.

Since aquifers of similar material behave more or less the same hydraulically, regardless of origin, why worry about genesis? We are interested in the origin of an aquifer because it provides a unique set of physical properties beyond porosity and permeability. These include textural variability, thickness, continuity, areal extent, geometry, orientation, and type of associated strata or materials. Because these qualities can greatly influence the conceptual hydrogeologic model we formulate, it is useful to consider the aquifer material's origin. For example, sandstone is a common aquifer material, but its role in the hydrologic system may differ if it was formed along a beach, on an alluvial fan, or in a dune field. This difference is not due to the origin itself, but due to what that origin provided in terms of sorting of grains, geometry, and associated deposits (Table 4-1). Recognizing the role these parameters play in controlling hydrologic behavior is an essential part of formulating a conceptual hydrogeologic model.

REFERENCES

Bathurst, R. G. C. 1975. *Carbonate sediments and their diagenesis.* Vol. 12. *Developments in Sedimentology.* New York: Elsevier. 658 p.

Driscoll, F. G. 1986. *Groundwater and wells.* St. Paul, MN: Johnson Division. 1089 p.

Fetter, C. W. 1994. *Applied hydrogeology.* Upper Saddle River, NJ: Prentice-Hall. 691 p.

Freeze, R. A., and J. A. Cherry. 1979. *Groundwater.* Englewood Cliffs, NJ: Prentice-Hall. 604 p.

Heath, R. C. 1983. *Basic ground-water hydrology.* Water-supply paper 2220. U.S. Geological Survey. 85 p.

Laney, R. L., and C. B. Davidson. 1986. Aquifer-nomenclature guidelines. Open-file report 86-534. U.S. Geological Survey. 46 p.

Stone, W. J., and B. E. McGurk. 1987. Hydrogeologic considerations in mining, Nations Draw area, Salt Lake coal field, New Mexico. Bulletin 121: 73–78. New Mexico Bureau of Mines and Mineral Resources.

Meinzer, O. E. 1923. The occurrence of ground water in the United States—with a discussion of principles. Water-supply paper 489. U.S. Geological Survey. 321 p.

Miller, J. A., and R. A. Renken. 1988. Nomenclature of regional hydrogeologic units of the southeastern coastal plain aquifer system. Water-resources investigations report 87-4202. U.S. Geological Survey. 21 p.

Seaber, P. R. 1986. Evaluation of classification and nomenclature of hydrostratigraphic units. *EOS Transactions of the American Geophysical Union* 67(16): 281.

Seaber, P. R. 1988. *Hydrostratigraphic units*. Vol. 0–2. In *Geology of North America—Hydrogeology*. W. Back, J. S. Rosenshein, and P. R. Seaber, eds. Boulder, CO: Geological Society of America.

PART II

THE HYDROLOGIC SYSTEM

Once you understand the geologic setting, the occurrence, movement, and quality of the water within it must be determined. As for the geologic setting, this involves compiling available information, characterizing the hydrologic conditions, and determining the hydrologic impact of the geologic setting.

THOUSAND SPRINGS, SNAKE RIVER CANYON, IDAHO.

The water issues from the open-textured part of a lava sheet. The height of the falls is 180 feet. The springs yield enough water to supply the city of New York. Photograph by I. C. Russell.

CHAPTER 5

Compiling Hydrologic Information

In addition to the geologic setting, a sound hydrogeologic study demonstrates a thorough understanding and, if possible, includes a quantification of the hydrologic system. Thus, after geologic information has been collected and synthesized, available hydrologic information must be compiled. As in the search for geologic information, you must know what hydrologic information to compile, the regional hydrogeologic province within which the study area lies, and the sources of hydrologic data, as well as some basic tools for presenting the information.

HYDROLOGIC INFORMATION

Various kinds of hydrologic data are needed in hydrogeologic studies. By searching literature, locate the major references on the study area's hydrology, or at least on the region within which it lies. Data are needed on the basic characteristics of each component of the hydrologic system: surface water, soil water, and ground water. Any information that permits the calculation of a water budget is of special interest.

Surface Water

It is important to include basic data on flowing and standing surface water in the area; surface water is one of the main targets in the compilation of hydrologic information. For streams, obtain information on the

- major drainage system(s)
- flow rate
  - (maximum, minimum, and average discharge, as well as flow-duration curves)
- water quality
  - general chemistry and field parameters (temperature, pH, specific conductance, turbidity, alkalinity, nitrate content, etc.)
  - dissolved solids (total and major-ion content)
  - metals (total and dissolved)
  - radionuclides
  - organic constituents
  - biological parameters (dissolved oxygen, biological oxygen demand, and taxa present).

For lakes, wetlands, and swamps, include information on

areal extent
volume of water contained
source of the water
water quality (as for streams)

When these parameters vary due to time and interaction of the surface water with soil and ground water, report the variations, as well.

Soil Water

Although less likely to be available, any information on conditions in the unsaturated zone is also useful. This is especially true when remediating contaminant spills or leaks, since the vadose zone is the first place where they may be contained or recovered. There is usually little published information about soil water and, if critical to the study, making field measurements will be necessary. Data to compile include

moisture content (range, average)
suction
rate of water movement (hydraulic conductivity)
chemistry of soil water

As for surface water, any information on the temporal or spatial variation in these parameters is also useful. Lacking specific data, you may use standard soils maps to estimate infiltration properties.

Ground Water

The most important component of the hydrologic system in most hydrogeologic studies is ground water. Data are needed on three main aspects of this water in the saturated zone: its occurrence, movement, and quality.

Information pertaining to ground-water occurrence falls into two main categories:

- major aquifer(s)
  - material type
  - extent
  - depth
  - thickness
  - hydrologic properties (yield, hydraulic conductivity, transmissivity, etc.)
- water level
  - depth
  - elevation
  - perched or regional

Also, determine whether the water is unconfined or confined.

For ground-water movement, provide information on

- recharge
  - source
  - area
  - rate
- flow
  - direction
  - rate
- discharge
  - area
  - rate

The means by which recharge and discharge are accomplished are also important, if known.

The information needed for water-quality is the same as that compiled for surface water, except for the biological parameters:

- general chemistry and field parameters (temperature, pH, specific conductance, turbidity, alkalinity, nitrate content, etc.)
- dissolved solids (total and major-ion content)
- metals (total and dissolved)
- radionuclides
- organic constituents

Long-term averages, as well as variation in these parameters through time, are of special interest. In ground-water contamination studies, background values from off-site are also needed.

HYDROLOGIC PROVINCES

Much has been written about the geographic extent or large-scale aspects of ground-water occurrence (for example, see Fuller 1905; Ries and Watson 1914; Meinzer 1923; Thomas 1954). Heath (1982, 1984) gave an excellent summary and comparison of previous attempts to delineate ground-water regions in the United States, as well as his own approach to classifying these ground-water systems. Most introductory hydrology textbooks have a section on ground-water regions, especially those recognized in the United States (Walton 1970; Todd 1980; Fetter 1994; Driscoll 1986). The number of provinces recognized ranges from 8 (Ries and Watson 1914) to 21 (Meinzer 1923). When working in a region for the first time, you should consult one of the references by Heath (1982, 1984).

As might be expected, the ground-water regions shown in Figure 5-1 are similar to the geologic provinces described in Chapter 2 (Figure 2-1). The uniformity of geologic conditions in these regions makes for uniform hydrogeologic conditions there as well.

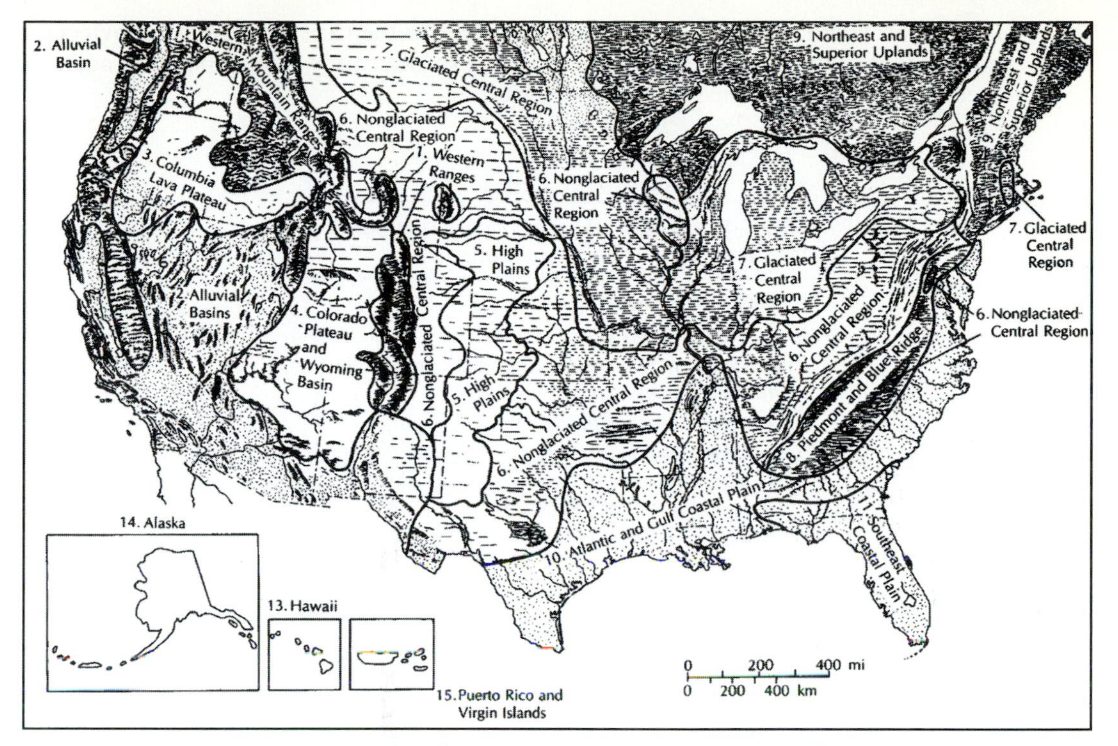

FIGURE 5-1 Hydrologic provinces in the United States. (From Fetter 1994, Figure 9.39.)

Ground-water regions over which there has been general agreement among the various workers on this topic include the Atlantic and Gulf Coastal Plain, New England, the Glaciated Upper Midwest, the High Plains, the Colorado Plateau, the Basin and Range country, the Western Mountains, and the Columbia Plateau.

Regardless of how many divisions are recognized or how they are defined, the significant point for hydrogeologists is that hydrologic conditions can be similar throughout broad areas. More specifically, Heath (1984) found five general factors to be consistent over large regions and used them to delineate his ground-water provinces:

components of the system
nature of water-bearing openings
composition of the rock matrix
storage and transmissivity characteristics
recharge and discharge conditions

An example of the kind of information available for provinces is given in Table 5-1. As noted by Heath, if hydrologic regions have been correctly identified, experience gained in one part should apply to every other part. As was the case for geologic provinces, if no previous hydrogeologic works are available for the study area, reports on adjacent areas in the same province should be relevant.

TABLE 5-1 Example of Generalizations That Can Be Made for a Given Hydrogeologic Province

Province—High Plains
Ground-Water Occurrence
unconfined aquifers—dominant; single, unconsolidated deposit; porosity large (> 0.2)
confined aquifers—insignificant
Ground-Water Movement
Recharge—by losing streams
Transmissivity—large (> 2,500 m^2/d)
Discharge—by contact and depression springs
Ground-Water Quality
Generally good; aquifers insoluble

Source: Data from Heath, 1984, Table 5.

SOURCES OF HYDROLOGIC DATA

In addition to the general characteristics of the hydrologic province involved, one needs detailed information on the study area itself. However, before setting out to collect new hydrologic data, it is wise to determine what is already known. Hydrologic information is available from many of the same sources listed for geologic information (see Chapter 2), as well as some additional ones.

National Government Agencies

In the United States, the main source of hydrologic information is the USGS—Water Resources Division. As noted in Chapter 2, they maintain a district office in a major city, if not the capital, of every state. These offices will have surface- and ground-water reports as well as well-record and water-quality files; they also provide access to their various national databases. If they don't sell copies of an item of interest, they will at least have an address where it can be ordered or, perhaps, a library where it may be reviewed.

Other federal agencies in the United States may have useful reports or information as well. These include the Army Corps of Engineers, Bureau of Land Management, Bureau of Reclamation, Department of Agriculture (especially agricultural experiment stations), Environmental Protection Agency, Forest Service, and the National Park Service. The government section of the telephone directory will give the location of the nearest office for such agencies. If an agency or office does not have the data needed, they should be able to suggest who does.

Outside the United States, similar national agencies are useful sources of hydrologic information. However, as noted in Chapter 2, the state agencies may have primacy over the national ones, unless a project spans state boundaries. Thus, in Australia, for example, the Bureau of Mineral Resources (the national geological survey) has produced the major references on such important hydrologic features as the Great Artesian basin, the Murray basin, etc.

State Agencies

Usually, the next best source of hydrologic information is state agencies. The state geological survey often produces its own water-resource reports or distributes such reports prepared in cooperation with other national or state agencies. Since various reports other than those designated as ground-water reports may also contain hydrologic data, some state surveys provide a booklet listing the water-resource information contained in their entire range of publications (e.g., that for New Mexico, by Stone 1992).

Files in the petroleum section of state surveys may also contain useful hydrologic information. For example, scout cards often include results of porosity and permeability tests for intervals cored, as well as the volume and quality of water recovered in drill-stem tests. Sometimes a separate file is kept of oil and gas wells that produced only water. If the water is usable, such wells are often turned over to the land owner. However, because of their unorthodox origin, they may not show up in traditional databases. Nonetheless, such wells provide additional opportunities to access the ground water in the area. In rare cases, such wells have neither been used nor properly abandoned. Thus, depending on their condition and local regulations, they may be converted into a water-supply well.

Other state agencies may also produce hydrologic reports or have useful hydrologic information on file. In the United States, these agencies include the regulatory agencies that oversee water rights (office of the state engineer), environmental protection (state environment department), petroleum production (state oil and gas commission), and mining activities (department of mines). They should also be contacted during the search for information, especially if data obtained from the major sources are inadequate. Locating relevant documents should not be difficult, since copies of the most commonly referenced sister-agency reports and selected environmental impact statements are often on file with the state geological survey.

Outside the United States, hydrologic data may be gathered by state agencies other than the geological survey as well. In South Australia, for example, the state Department of Engineering and Water Supply conducts and funds water-resource studies. Similar state agencies no doubt exist in other countries as well.

Universities

As in the case of geologic information, university libraries are also good sources of hydrologic data. More specifically, there may be hydrologic theses or dissertations filed there that pertain to your study area or problem. Sometimes bibliographies of water-related theses and dissertations for a given university are available (for example, that for New Mexico Institute of Mining and Technology by Myers and Herman 1982). University libraries are often the official repository where the public may review environmental impact statements or other environmental documents produced by various facilities. Of course, such libraries are also the best place to find a wide range of technical journals, including those on hydrology.

Professional Organizations and Journals

Professional hydrologic organizations are also an important source of hydrologic information. Their conferences and symposia may include presentations on topics pertinent to your project. Some of the organizations in the United States hold regional conferences as well as national ones. The meetings in your region are especially likely to include hydrologic conditions and problems similar to those in your study area. There are often state chapters of some of these organizations. Proceedings of their annual meetings may also include relevant papers or additional data sources. Although not strictly so, those organizations based in the United States (such as the American Geophysical Union, the American Water Resources Association, the National Ground Water Association, etc.) tend to cover mainly North American studies, whereas those based overseas or that target international audiences (such as the International Association of Hydrogeologists, the International Water Resources Association, etc.) have a more global coverage.

The journals of these professional organizations (such as the *Water Resources Bulletin, Ground Water,* etc.) may contain papers that cover your study area or solve a hydrologic problem related to your own. Journals not affiliated with any professional organization (i.e., *Journal of Hydrology, Journal of Contaminant Hydrology,* etc.) may also contain articles with useful information for your study.

Consulting Reports

Although less visible than agency publications or journal articles, reports generated by consulting companies also contain valuable hydrologic data. Consulting reports on hydrologic studies are generally more abundant and accessible than those on geologic investigations. This is because geologic consulting reports are commonly associated with mineral-resource or petroleum exploration and, thus, tend to be confidential. Once a consulting report has been submitted to the state in conjunction with a regulatory action, such as a request for water rights or an environmental clean-up case, it is public information. Such reports are filed by the appropriate state agency's office, where they may be viewed or copied. To determine the availability of consulting reports for your study area contact either a company known to be working in the region or the state agency most likely to be involved with the problem. Although the state geological survey is usually not a regulatory agency, they often also file selected consulting reports, especially if they are the only source of information on an area or deal with a topic of current public and professional interest.

Miscellaneous Sources

Various other entities may also be sources of hydrologic information. For example, the National Academy of Sciences (1974) investigated water conservation and harvesting in arid regions. In some developing nations, international organizations (such as the World Health Organization) are an important source of water-resource information. The World Bank has also addressed water problems in various parts of the world, such as village water supplies in developing nations (Saunders and Warford 1976). Although these studies may have been done in one area, their results may be applied to others.

Internet

Countless Internet sites are devoted to hydrologic topics, and this is an important new source of information. Some sites are associated with water-resource concerns in a geographic region. Others tout the information and services available from various government agencies. Still others are focused on a given hydrologic topic, such as ground-water modeling. For example, there are several sites for just the USGS code MODFLOW.

HYDROLOGIC TOOLS

As with the geologic setting, there are various tools for compiling, portraying, and eventually characterizing the hydrologic system. These include both raw data and interpretive illustrations. Hydrologic tools may be separated into three categories based on the part of the hydrologic cycle that is involved: surface-water tools, soil-water tools, and ground-water tools.

Surface-Water Tools

Various tools are useful in compiling and illustrating surface-water information. These include tables, graphs, maps, photographs, and cross sections.

Tables generated by government agencies are a basic tool in compiling surface-water data. The most common type of tabular information available for surface water is stream-gage data. These tables include such information as location, contributing drainage area, period of record, type of gage, and remarks as to the quality of the record, construction details, etc. If the gage is on a river, the stage or discharge values will be presented. If located on a reservoir, data will be in acre-ft. Values for total, maximum, minimum, and mean are normally included. The annual reports of the USGS—Water Resources Division are not only a good source of surface-water data, but also give good examples of what can be presented and how (Table 5-2).

TABLE 5-2 Annual Report Rio Grande Basin 08266820 Red River Below Fish Hatchery, Near Questa, NM

Discharge, Cubic Feet per Second, Water Year October 1992 to September 1993 Daily Mean Values

Day	Oct	Nov	Dec	Jan	Feb	Mar	Apr	May	Jun	Jul	Aug	Sep
1	59	48	e47	60	51	54	57	153	391	194	77	128
2	58	47	49	60	50	51	59	148	405	185	82	119
3	57	45	50	59	49	50	61	141	392	176	77	113
4	55	44	50	49	50	50	58	132	356	173	85	105
5	54	44	50	47	45	48	60	136	332	171	92	99
6	53	44	50	57	43	51	61	144	312	165	95	97
7	52	45	44	61	45	53	61	142	282	157	90	97
8	51	47	48	65	47	55	60	139	257	146	85	96
9	50	47	51	62	47	56	60	131	239	142	84	93
10	49	48	53	60	47	55	61	122	219	136	84	88
11	46	49	54	61	46	54	62	120	197	132	80	86
12	44	47	55	55	46	52	65	125	194	131	77	82
13	42	47	54	52	46	47	70	146	200	132	82	82
14	e46	48	46	59	46	49	77	188	228	126	100	95
15	e48	48	49	58	47	48	79	211	254	123	98	94
16	49	49	46	56	46	49	79	237	289	117	88	88
17	53	49	47	58	47	50	79	283	308	114	83	85
18	50	49	54	56	45	51	70	284	306	112	81	81
19	50	50	56	56	52	52	71	277	277	106	87	78
20	48	51	48	55	70	52	76	288	265	106	97	77
21	48	53	49	53	63	52	75	310	269	104	92	76
22	49	49	49	54	52	53	83	341	269	96	105	75
23	51	52	47	54	49	55	96	358	255	93	93	74
24	52	50	48	47	53	59	107	346	243	91	87	73
25	50	46	49	45	53	59	102	340	232	89	84	72
26	50	44	47	48	53	58	102	413	223	86	81	71
27	48	43	46	50	53	63	119	437	211	84	104	70
28	50	42	52	49	54	58	129	426	200	80	159	68
29	50	43	61	51	—	59	135	396	194	74	152	67
30	48	e45	65	51	—	58	147	376	195	73	135	66
31	51	—	62	51	—	57	—	374	—	74	139	—
Total	1561	1413	1576	1699	1395	1658	2421	7664	7994	3788	2955	2595
Mean	50.4	47.1	50.8	54.8	49.8	53.5	80.7	247	266	122	95.3	86.5
Maximum	59	53	65	65	70	63	147	437	405	194	159	128
Minimum	42	42	44	45	43	47	57	120	194	73	77	66
Ac-ft	3100	2800	3130	3370	2770	3290	4800	15200	15860	7510	5860	5150
Statistics of Monthly Mean Data for Water Years 1978–1993, By Water Year (WY)												
Mean	53.7	46.7	42.9	43.4	43.9	48.0	82.3	201	219	106	72.5	61.9
Maximum	71.0	59.2	51.0	55.3	57.9	72.0	144	368	520	226	95.3	86.9
(WY)	1986	1992	1987	1992	1992	1989	1985	1985	1979	1979	1993	1986
Minimum	29.0	33.0	28.2	31.4	31.5	35.1	39.7	50.5	56.8	43.1	42.1	31.2
(WY)	1979	1979	1979	1979	1981	1981	1981	1981	1981	1981	1981	1978

Summary Statistics	For 1992 Calendar Year		For 1993 Water Year		Water Years 1978–1993	
Annual Total	30026		36719			
Annual Mean	82.0		101		85.5	
Highest Annual Mean					129	1979
Lowest Annual Mean					41.9	1981
Highest Daily Mean	213	May 2	437	May 27	676	May 27 1979
Lowest Daily Mean	42	Oct 13	42	Oct 13	26	Oct 10 1978
Annual Seven-day Minimum	44	Nov 25	44	Nov 25	26	Dec 9 1978
Instantaneous Peak Flow			466	May 27	755	Jun 8 1979
Instantaneous Peak Stage			4.04	May 27	5.30	Jun 8 1979
Instantaneous Low Flow			36	Nov 4	21	Dec 14 1986
Annual Runoff (Ac-ft)	59560		72830		61920	
10% Exceeds	176		230		169	
50% Exceeds	59		61		55	
90% Exceeds	48		47		37	

e Estimated

Location—Lat 36°40'54", long 105°39'21", in NW¼NW¼ sec. 10, T.28 N., R.12 E., Taos County, Hydrologic Unit 13020101, on right bank 0.3 mi downstream from State Fish Hatchery, 3.5 mi upstream from mouth, and 3.7 mi southwest of Questa.

Drainage area—185 mi^2.

Period of Record—August 1969 to July 1978 (discharge measurements only), August 1978 to current year.

Gage—Water-stage recorder. Elevation of gage is 7,070 ft above National Geodetic Vertical Datum of 1929, from topographic map. Prior to Aug. 16, 1979, at site about 250 ft upstream at datum 5.55 ft higher.

Remarks—Records good except for estimated daily discharges, which are poor. Diversions for irrigation of about 3,000 acres upstream from station. Several observations of water temperature were made during the year.

Source: From Cruz and others 1994.

Graphs are another important tool in surface-water hydrology. They show the relationship between two variables in x,y space. Common examples are the hydrograph and the flow-frequency curve. Hydrographs can be prepared to plot various parameters against time. This may include water stage (height) or discharge (flow) of a stream or stage or storage for a reservoir. Such graphs are probably the most commonly used tool in surface-water hydrology. They quickly summarize and illustrate the specific data or variation thereof for any time period. Flow-frequency curves show the flow that can be expected a certain percent of the time. For example, by using such graphs you may learn the flow that will be equaled or exceeded 10%, 50%, 99%, or any other percent of the time. Graphs are also the best way to show the variation through time (or with position, in the case of a stream) of selected water-quality parameters (salinity, concentration of a specific constituent, temperature, pH, etc.) at a given point on a stream or reservoir.

Maps are an obvious tool for determining and presenting surface-water information, due to their spatial characteristics. Because streams and lakes are major features of the landscape, they are already shown on standard topographic maps. This makes the illustration of several surface-water characteristics easier. For example, the area that would be flooded at different river stages or by various standard floods (10-year, 100-year, etc.) can easily be shown by the addition of lines indicating projected high-water marks to topographic sheets or topographic-sheet overlays. Harrison (1968) made maps showing the areas to be flooded at various stages of the Red River around Grand Forks, North Dakota. Watershed boundaries, drainage nets, tributaries of interest, paleochannel position(s), and channel change may be shown in a similar manner. For example, Maulde and Scott (1977) mapped headcut position at different times to show the rate of arroyo development near Santa Fe, New Mexico.

Various remote-sensing products are useful tools as well. This includes both satellite and aircraft imagery. Color, black-and-white, color-infrared or infrared coverage may be available for your area. Sources include not only the federal outlets (e.g., NASA) but also institutes affiliated with local universities.

Many of the phenomena previously mentioned could also be illustrated on standard aerial or satellite photographs. Aerial photographs are especially useful in mapping. For example, air photos may be used to show terraces along streams, changes in the surface area of lakes and ponds, the extent of wetlands and swamps, the distribution of vegetation, or contaminants around reservoirs, etc.

The impact of dams on rivers may be shown by before-and-after or sequential ground-based photographs. For example, Keller (1992) used photos taken from the same place at different times to emphasize changes in the Colorado River due to the construction of the Glen Canyon Dam, Arizona. More specifically, these photos documented that the river bed was aggrading due to the build-up of flash-flood detritus from tributaries. These deposits had always been supplied to the river, but the reduced flow of the Colorado, due to the dam, was insufficient to permit flushing them away.

Profiles and cross sections are also essential tools in documenting stream characteristics. Longitudinal profiles can be used to show such things as gradient, knick points, and local base levels. Transverse sections, constructed for the same place at different times, are useful in showing changes in the channel due to scour and fill (Figure 5-2). If the water table is included on such illustrations, the stream and ground-water relationship may also be shown. This is of special interest in arid regions, where the two systems are not necessarily in direct contact.

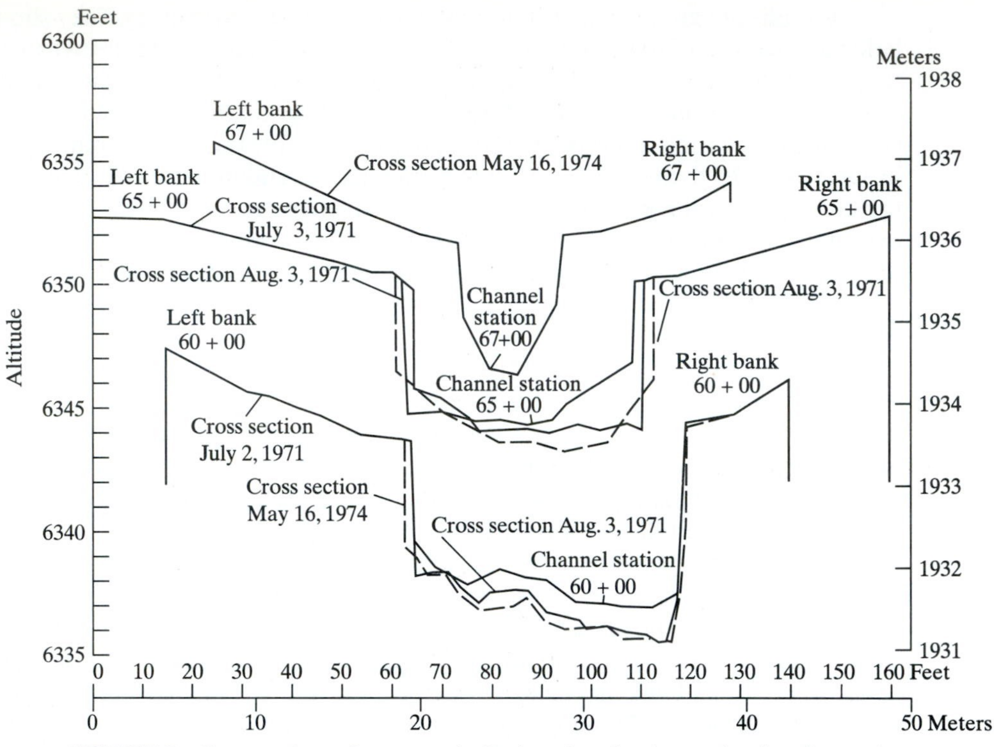

FIGURE 5-2 Cross sections of an arroyo in the American Southwest, showing changes in channel area between 1971 and 1974. (From Observations of contemporary arroyo cutting near Santa Fe, New Mexico USA by Maulde and Scott, copyright 1977 by John Wiley & Sons Limited. Reproduced with permission.)

Soil-Water Tools

The compilation and portrayal of vadose-zone conditions is also enhanced by the use of various tools. These include tables, maps, profiles, and cross sections.

Various types of soil-water data may be presented in tables. These data include water content (gravimetric or volumetric), percent saturation, hydraulic conductivity (unsaturated and saturated), moisture retention, matric suction, bulk density, porosity, etc. Such tables should be presented to support any diagrams based on them.

Areal variation in soil-water phenomena may be expressed by various kinds of maps. For example, the distribution of physical parameters (such as water content, grain size, bulk density, etc.) may be readily conveyed by maps. Similarly, the soil-water chemistry (the concentration of a given constituent at a given depth, the average concentration over a given depth interval, or the maximum concentration regardless of depth) may also be presented in map form. The variation in such physical and chemical parameters may be shown either by contours or by zones for which the range of values is labeled directly or indicated by a pattern explained in a legend.

Maps may also be used to report the results of geophysical surveys of the vadose zone. More specifically, surface electromagnetic measurements have been used to map geology, shallow subsurface conditions, soil salinity, and contaminant plumes. For example, Walker and others (1990) used frequency-domain electromagnetic-induction techniques to supplement drilling data to determine the nature of the unsaturated zone in southeastern South Australia. Two different devices (Geonics EM 31 and EM 34) with various combinations of dipole configuration (horizontal and vertical) and coil spacing (3.7, 10, 20, and 40 m) were used to determine the depth of the boundary between layers in the unsaturated zone (Figure 5-3).

Profiles are commonly used to illustrate the vertical distribution of vadose-zone constituents. These profiles are constructed by plotting analytical data versus depth. For

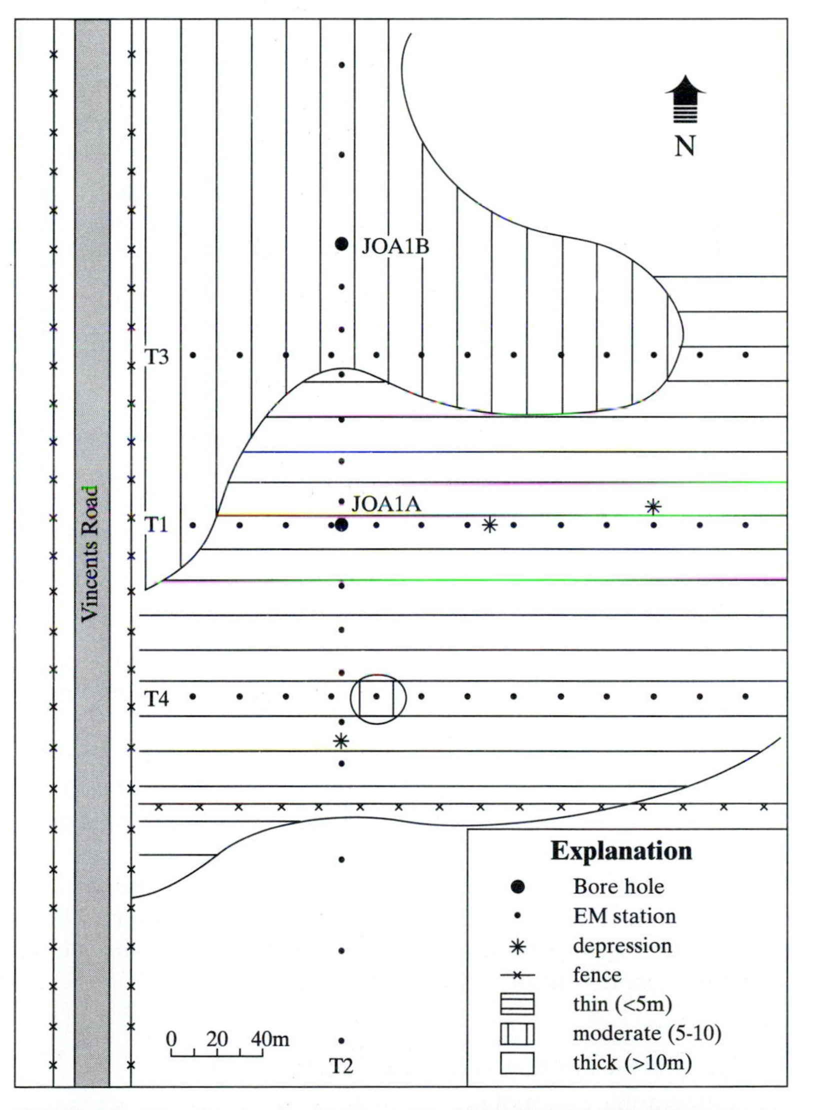

FIGURE 5-3 Map showing the thickness of the high soil-water chloride layer in the unsaturated zone near Joanna, South Australia, based on electromagnetic surveys along transects T1-T4. (From Walker and others 1990, Figure 22.)

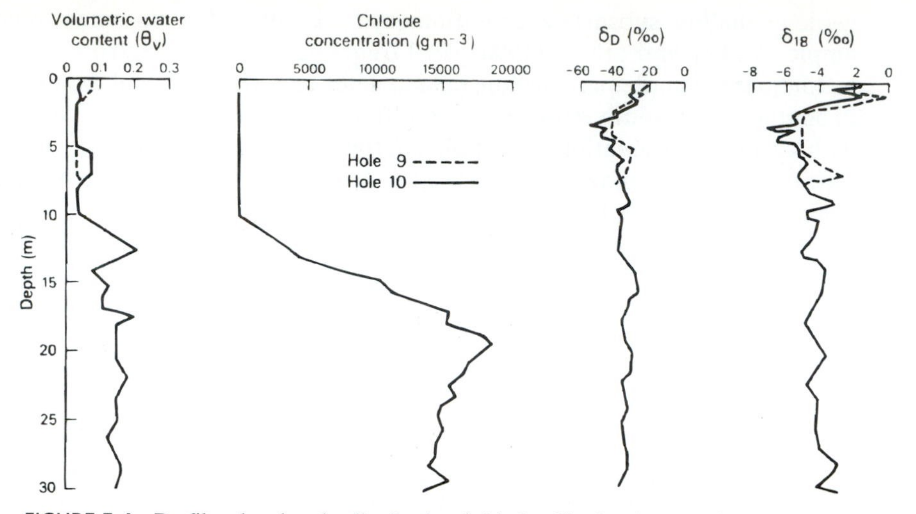

FIGURE 5-4 Profiles showing the distribution (with depth) of moisture, soil-water chloride, and the stable isotopes of oxygen and hydrogen in a portion of the unsaturated zone beneath a dune from which native vegetation has been cleared. Area located in the Murray Basin, South Australia. (Reprinted from Journal of Hydrology, v. 76. Allison, Stone and Hughes, Recharge in karst and dune elements of a semiarid landscape as indicated by natural isotopes and chloride, p. 1–25, 1985, with permission from Elsevier Science.)

example, moisture content or the concentration of specific dissolved constituents in the soil water may be presented in this manner (Figure 5-4).

The relationship between adjacent observations of the vadose zone may best be shown in a cross section. Such sections may be constructed using logs of auger holes, concentration and depth profiles, or the results of electromagnetic surveys.

Ground-Water Tools

To compile and clearly illustrate ground-water phenomena, you may use a variety of tools. Most commonly used are tables, well logs, cross sections, maps, and various types of water-quality diagrams.

Perhaps the most important tools are tables, because they support all the other tools. Without them, maps, cross sections, and other diagrams could not be created, let alone evaluated. Two main types of tables are used to present ground-water information: those giving well data and those giving water-quality data.

All the basic information on wells is included in a table that has traditionally been referred to in USGS and other reports as the "records of wells" (Table 5-3). This is a well inventory. Different people include different information in such tables, but the essential parameters to include in well-record tables are

well name
location (subsection, section, township, range)
topographic quadrangle
date drilled

TABLE 5-3 Records of Selected Wells in Quay and Adjoining Counties, New Mexico*

Location No.	Owner or Name	Year Completed	Depth (feet)	Diameter (inches)	Altitude (feet)	Water level: Depth Below Land Surface (feet)	Water level: Date	Stratigraphic Unit	Type of Pump and Power Source	Use of Water	Remarks
5.26.22.320	Abercrombie and H Hawkins No. 1—Nappier	1949	7149	9	4518	—	—	—	—	—	Oil test; in DeBaca Co., 2½ miles west of Quay Co. line; log
5.27. 1.341	L. W. Barnhill	—	200	5	4950	131.0	8-23-55	To	P, W	D, S	—
3.312	—	—	77.1	5	4650	49.6	4-15-55	Qal, Ꞧ c	P, W, I	S	—
9.333	L. W. Barnhill	—	33.3	6	4530	28.5	8-25-53	Qal	P, W	S	Ca
12.444	L. W. Barnhill	1951	182	6	4920	170	1951	To	P, W	S	T 61° F
15.424	Dick Ballew	1945	64.8	4	4640	43.9	4-15-55	Qal, Ꞧ c	P, W	D, S	T 62° F, Ca Pumping water level
17.441	Mrs. N. G. Koll	—	7.6	48	4510	5.3	4-15-55	Qal, Ꞧ c	P, W	S	Dug. T 51° F
25.242	D. O. Bomar	Old	86	5	4890	75	1954	To	P, W	D, S	T 63° F
29.212	Guy Shipely	—	35.1	7	4470	25.7	4-15-55	Qal	P, W,	D, S	—
30.242	Mrs. N. G. Koll	—	13.3	48	4440	12.9	4-15-55	Qal, Ꞧ c	P, W	S	Dug. Pumping water level. Est yield 4 gpm
31.122	Mrs. N. G. Koll	—	13.2	30	4420	13.2	4-15-55	Qal	P, W	S	Dug. Est yield 1 gpm. T 58° F
5.28. 1.111	G. E. Murphy	1950	90	6	4750	70	1950	To	P, W	S	John Maddox, driller

*Location number: See text for explanation of well numbering system.

Year completed: Wells designated "old" drilled generally before 1925.

Depth: Depths are in feet below land surface. Reported depths are given to nearest foot. Measured depths are given to nearest tenth of a foot.

Diameter: The diameter of the casing, or the mean diameter of the well if uncased, to nearest inch.

Altitude: Altitude of land surface at well. Altitude interpolated from topographic maps or aneroid determination to nearest 10 feet.

Water level: Reported depths are given to nearest foot. Measured depths are given to nearest tenth of a foot.

Stratigraphic unit: Qal, younger alluvium, Qc, upland cover of older alluvium; To, Ogallala Formation; Ks, Cretaceous sandstone and siltstone; Jm, Morrison Formation; Je, Entrada Sandstone; Ꞧ c, Chinle Formation; Ꞧ sr, Santa Rosa Sandstone; Pr, Permian rocks.

Type of pump and power source: E, electric; I, internal combustion; J, jet; N, none; P, plunger or cylinder; S, submersible; T, turbine; W, windmill.

Use of Water: D, domestic; I, irrigation; Ind, industrial; O, observations; PS, public supply; RR, railroad; S, stock; N, none.

Remarks: All wells are drilled and cased with steel casing unless otherwise indicated. Ca, chemical analysis in table 3; dd, drawdown; est, estimated; gpm, gallons per minute; log, log in table 6; meas, measurement; perf, perforated, perforations given in feet below land surface; rept, reported, reportedly; T 61° F, temperature in degrees Fahrenheit; USBR, U.S. Bureau of Reclamation; yields are reported unless otherwise indicated.

Source: From Berkstresser and Mourant 1966.

total depth
aquifer
ground-surface elevation
water-level depth
water-level date
water-level elevation
screen specifications
screened interval
filter-pack specifications
filter-packed interval
aquifer properties
well use
availability of chemical analyses (Y/N)
log(s) available (type, depth interval)
source of information (reference for published sources)

The well-construction information is very important. For example, knowing that a well has a long screened interval or multiple screened intervals is critical when interpreting the associated water level or water-quality data. Even though well-construction diagrams are available, the information should be tabulated as well. If it does not readily fit in the well-records table, it should be presented separately.

Observations on springs are given in a table called "records of springs" (Table 5-4). Usually, several types of data are included.

spring name
location
topographic quadrangle
elevation
aquifer
spring type (geologic control)
flow (gpm)
flow date
field water-quality measurements (general chemistry)
development (if any)
use (stock watering, wildlife, etc.)

The field chemistry data are especially important if samples are not taken for lab analysis, because they are the only indication of water quality. Other information, such as seasonal variation, may be added in a "comments" column.

The other tables used deal with water-quality. These give results of chemical analyses of ground water from wells and springs. Separate tables are usually presented for general chemistry (Table 5-5), major ions (Table 5-6), metals (Table 5-7), radionuclides

TABLE 5-4 Records of Springs in Quay County, New Mexico*

Location number: See explanation in text.

Altitude: Altitude of land surface at spring. Altitude interpolated from topographic maps or aneroid determination to nearest 10 feet.

Stratigraphic unit: Qal, younger alluvium; Qc, upland cover of older alluvium; To, Ogallala Formation; Ks, Cretaceous sandstone and shale; Je, Entrada Sandstone; ꞦR c, Chinle Formation, ꞦR sr, Santa Rosa Sandstone.

Location Number	Owner	Name	Topographic Situation	Altitude (feet)	Stratigraphic Unit	Yield (estimated gpm)	Date	Use of Water	Temperature (° F)	Remarks
7.30.15.432	—	—	Below cliff in gully	4720	To	Seep	8-25-53	None	—	Reported good quality and to have supplied 25 families 1910 to 1930
8.27. 6.430	H. G. Johnson	—	Side of cliff	5100	Je	2	11-2-55	Stock	—	Perched water, piped to tank
8.31.12.320	—	—	Stream channel	4220	Qal	2	4-21-55	—	—	—
8.32.18.223	—	—	Stream channel	4220	Qal	5	4-16-55	Stock	—	—
35.114	Elder Dennis	—	Stream channel	4480	Ks	5	4-2-55	None	—	Spring at fault contact of Cretaceous and Triassic rocks
9.27.36.244	Mr. Hortenstein	Louisiana Spring	Side of cliff	5220	Ks	2	10-27-53	Stock	55	Chemical analysis in Table 3
9.32.24.322	Mrs. Hut Wallace	—	Stream channel	4200	Qal	1	4-8-55	Stock	58	—
33.333	S. S. Hodges	—	Stream channel	4190	Qal	25	4-16-55	Stock	—	—
9.33.24.312	Mr. Pierce	Hopkins Spring	Stream channel	4480	Ks	Seep	2-14-55	None	—	—
10.33.14.212	Mr. Stams	Starns Spring	Side of cliff	4080	Qc	Seep	2-15-55	Stock	—	—
10.35.32.422	Chapman Bros.	—	Stream channel	4020	Qal	3	12-1-54	Stock	—	Piped to tank
10.36. 8.233	Chapman Bros.	—	Steep slope	3920	ꞦR c	3	11-29-54	Domestic and stock	—	Piped to tank
18.224	Chapman Bros.	—	Stream channel	3970	ꞦR c	1	11-29-54	Stock	—	—
11.33.29.211	Otto Collins	—	Side of cliff	3920	ꞦR c	0.5	2-18-55	Domestic and stock	51	Piped to tank
11.36.30.412	Grady Oldham estate	—	Steep slope	3950	ꞦR c	0.5	11-5-54	Stock	—	—

Source: From Berkstresser and Mourant 1966.

TABLE 5-5 Miscellaneous Water-Quality Parameters for Ground Waters, Carlin Trend

Total

Well	Date Sampled	pH (F) SU	pH SU	SC (F) uS/cm	SC (L) uS/cm	Alt	Hard	T (C)	Turb SU	Color SU	Lang Index	DO	1DS @180 C	ISS @105 C	CN (T)	CN (F)	CN (W)
CS–2	02-Jul-91		7.6		553	272	294		17	10			462	4	0.007	–0.005	–0.005
I1	18-Oct-90		8.1		1600	230	880		30				1100	2900	–0.005*	–0.1	–0.005
I2	20-Oct-90		8.1		690	190	240		36				450	96	–0.005	–0.1	–0.005
I3	20-Oct-90		8.1		420	160	180		27				270	78	–0.005	–0.1	–0.005
I4	20-Oct-90		8.1		630	170	440		190				340	940	–0.005	–0.1	–0.005
I6	19-Oct-90		8.1		880	200	370		41				520	810	–0.005	–0.1	–0.005
J–2	22-May-91		8.1		384	197	211		37	20			328	84	–0.005	–0.005	–0.005
LJKC–1	10-Jul-91		7.6		693	219	324		2.4	10			576	3	0.014	–0.005	–0.005

Dissolved

Well	Date Sampled	pH (F) SU	pH SU	SC (F) uS/cm	SC (L) uS/cm	Alt	Hard	T (C)	Turb SU	Color SU	Lang Index	DO	1DS @180 C	ISS @105 C	CN (T)	CN (F)	CN (W)
G24	26-Aug-90	7.4	7.7		3/9	185		16	5			6.1	230			–0.00001	

Explanation
(F) = field
(L) = lab
SU = standard units
ALK = alkalinity (HCO3/CaCO3)
HARD = hardness (CaCO3)

T = temperature (Celsius)
Turb = turbidity
Lang Index = Langeller Index (observed pH – saturation pH; dimentionless)
DO = dissolved oxygen
TDS = total dissolved solids

TSS = total suspended solids
CN = cyanide
(T) = total
(F) = free
(W) = weak acid dissociable

Minus sign before value indicates less than detection limit given.

Source: From Stone and others 1991.

TABLE 5-6 Major-Ion Content (mg/L) for Ground Waters, Carlin Trend

Well	Date Sampled	Ca	Na	K	Mg	SO4	CI	CO3	HCO3	NO3 as N	NO2 as N	NO3/NO2	NH4 as N	NH4 as N Organic	F	Si	PO4 as P
Total																	
CS–2	02-Jul-91	87	7	4.5	21.6	58	8		272	–0.05	–0.01		–0.05		0.66	11.1	0.02
I1	18-Oct-90	190	39	14	100	310	31	–5	280	–0.05			–0.2		0.5	28	–0.02
I2	20-Oct-90	63	54	7.8	20	95	19	–5	230	–0.05			–0.2		–0.5	20	0.03
I3	20-Oct-90	48	13	3.7	15	40	8	–5	190	1.1			–0.2		–0.5	10	0.17
I4	20-Oct-90	110	29	13	37	80	13	–5	200	–0.05			–0.2		1.2	47	0.32
I5	19-Oct-90	90	22	7.5	36	150	15	–5	250	–0.05			–0.2		0.7	13	0.03
J–2	22-May-91	67	9.7	2.9	20.2	37	7			–0.1	–0.01		–0.05		0.29		0.22
LJKC–1	10-Jul-91	71	26	8	47.7	190	2			–0.05	–0.01		1.21		0.61	7.9	0.02
Dissolved																	
Well	**Date Sampled**	**Ca**	**Na**	**K**	**Mg**	**SO4**	**CI**	**CO3**	**HCO3**	**NO3 as N**	**NO2 as N**	**NO3/NO2**	**NH4 as N**	**NH4 as N Organic**	**F**	**Si**	**PO4 as P**
CS–2	02-Jul-91	133	7	4.5	22.6											9.8	
G24	28-Aug-90	34	24	2.1	18	6.3	7.3		226		–0.01	0.9	–0.01	–0.2	0.2	47	0.01
J–2	22-May-91		9.5	2.8	19.6										–0.05	20	
LJKC–1	10-Jul-91	83.9	27	8	48.3											6.8	

Source: From Stone and others 1991.

TABLE 5-7 Trace-Metal Content (mg/l) of Ground Waters, Carlin Trend

Total

Well	Date Sampled	Al	As	Ba	Be	B	Cd	Cr	Co	Cu	Au	Fe
CS–2	02-Jul-91	–0.1	–0.005	0.03	–0.005	–0.1	–0.005	–0.002	0.009	–0.005	0.014	0.74
I1	18-Oct-90	20	0.07	1.5		0.2	0.015	0.072		0.37	–0.005	55
I2	20-Oct-90	4.1	–0.005	0.06		0.2	–0.005	0.01		0.014	–0.005	4.7
I3	20-Oct-90	1.2	0.009	0.08		–0.1	–0.005	0.007		0.009	–0.005	4.9
I4	20-Oct-90	49	0.027	0.59		0.2	0.009	0.056		0.16	–0.005	41
I5	19-Oct-90	3.8	0.035	0.43		0.1	–0.005	0.015		0.09	–0.005	16
J–2	22-May-91	–0.1	0.006	–0.1	–0.005	0.59	0.013	0.013	–0.05	0.02	–0.05	1
LJKC–1	10-Jul-91	–0.1	0.01	0.03	–0.005	0.18	–0.005	–0.002	0.007	0.005	0.005	0.26

Dissolved

Well	Date Sampled	Al	As	Ba	Be	B	Cd	Cr	Co	Cu	Au	Fe
CS–2	02-Jul-91	–0.1	0.005	0.04	–0.005	0.1	–0.005	–0.002	0.009	0.005	0.012	0.89
G24	28-Aug-90	–0.01	0.007	0.1	–0.00005		–0.0001	0.003	–0.003	0.001		0.019
J–2	22-May-91	–0.1	0.005	–0.1	–0.005		0.009	–0.002	–0.05	–0.01	–0.05	–0.05
LJKC–1	10-Jul-91	–0.1	–0.005	0.02	–0.005	0.1	–0.005	–0.002	0.008	0.005	0.009	0.11

Source: From Stone and others 1991.

(Table 5-8, p. 75), and organic constituents (Table 5-9, p. 76). Not all may be necessary, depending on the site and study.

As the most basic information about ground-water occurrence comes from wells, a major tool for illustrating this information is the well log (Figure 5-5). As a minimum, logs should include the nature of the materials penetrated, total well depth, depth at which saturation was encountered, and static water level (if different). Ideally, well-construction information (screen size, screened interval, annular material, and intervals) is also shown. The log in Figure 5-5 would be ideal if the water level were also indicated; there appears to be plenty of room for this on the right side of the diagram.

Cross sections based on well logs are another good source of information. Such sections generally show stratigraphic units, folds, faults, static water level, and ground-water flow direction. These may be site-specific or generalized to show the conceptual hydrogeologic model for the area.

Various kinds of maps are useful in characterizing the spatial variation of ground-water phenomena. For example, the water level in an area is usually depicted by means of a contour map, using elevations of the water table or potentiometric surface, depending on whether unconfined or confined conditions prevail. If both occur, their water levels should be mapped separately. Aquifer properties (hydraulic conductivity, transmissivity, yield, specific capacity, etc.) and water quality (specific conductance, total-dissolved-solids content, sulfate content, etc.) may also be mapped. These parameters

Pb	Ll	Mn	Mo	Hg	Ni	Se	Si	Ag	Sr	Th	V	Zn
–0.002	–0.005	0.12	0.008	0.0003	–0.05	–0.005	11.1	–0.005	0.08	0.06	–0.005	0.073
0.049		0.29		0.0026	0.48	0.055	28	–0.005	0.67	–0.005		2.2
–0.005		0.12		0.0002	0.02	–0.005	20	–0.005	0.5	–0.005		0.03
0.005		0.2		0.0001	0.03	–0.005	10	–0.005	0.17	–0.005		0.11
0.035		2.2		0.0003	0.14	–0.005	47	–0.005	0.64	–0.005		0.43
0.029		0.17		0.0005	0.14	0.024	13	–0.005	0.31	–0.005		0.28
0.003	–0.005	0.1	–0.1	0.0003	–0.02	–0.005		–0.005	0.1		0.024	0.095
–0.002	0.02	0.07	0.84	0.004?	0.09	–0.005	7.9	–0.005	0.43		0.009	0.057

Pb	Ll	Mn	Mo	Hg	Ni	Se	Si	Ag	Sr	Th	V	Zn
0.002	–0.005	0.13	0.006	0.0002	–0.05	–0.005	9.8	–0.005	0.07		–0.005	0.008
–0.001	0.013		–0.01	0.0002	–0.0001	–0.001	47	–0.001	0.15		0.011	0.19
–0.002	–0.005	19.6	–0.1	–0.0002	–0.02	–0.005		–0.005	0.1		–0.005	0.026
–0.002	0.019	0.08	0.85	0.0027	0.08	–0.005	6.8	–0.005	0.46		–0.005	0.03

may be shown by contours or zones. If there are several aquifers in the area, separate maps should be prepared for each. Stone and others (1983) attempted to present a suite of three hydrologic maps for each sandstone aquifer in the San Juan basin of northwest New Mexico: one for water level, one for transmissivity, and one for specific conductance. Obviously, only those maps for which available information was sufficient could be constructed. An example of a complete suite of such maps for the Gallup Sandstone is given in Figure 5-6, p. 77.

Water chemistry may also be shown with various special diagrams. The major-ion chemistry of a ground-water sample is often easier to visualize or compare with that of other samples when it is plotted on Stiff, pie, or Piper diagrams (Figures 5-7 and 5-8, p. 78). Consult standard hydrology textbooks or the original references for these diagrams (given at end of chapter) for the steps used to prepare them. Stiff or pie diagrams may be placed on hydrogeologic maps for an even better means of showing the similarity or variation in water quality across an area.

Completeness

As with geologic tools, all hydrologic illustrations must be complete to be clear. This can be assured by fully labeling items, having a comprehensive explanation (legend), and using appropriate, stand-alone captions. On graphs, clearly indicate parameters as-

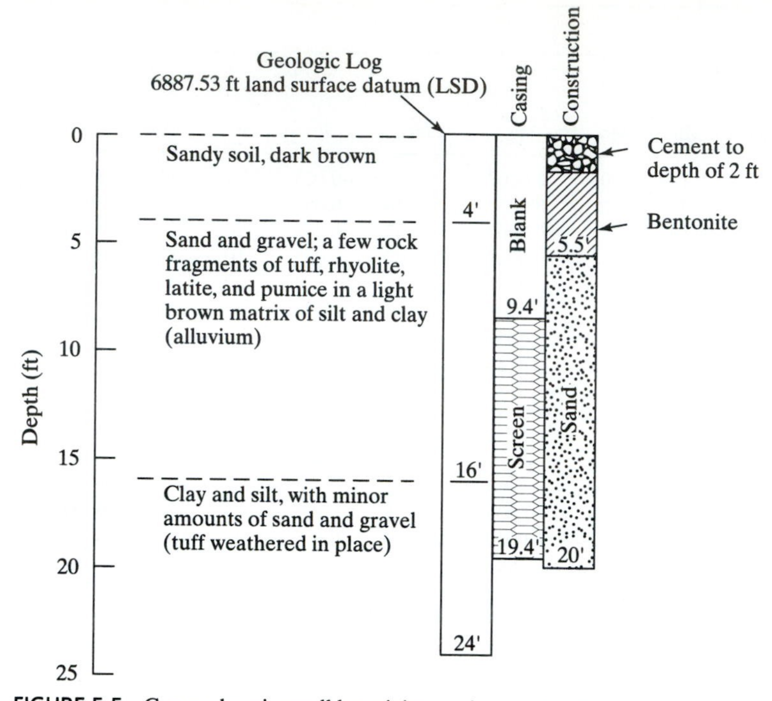

FIGURE 5-5 Comprehensive well log, giving geology and well-construction details. (From Purtymun 1995, Figure VIII-K.)

signed to the various axes, as well as the units of measure used. Further suggestions on illustrations are given in Chapter 9.

In monitoring reports that present ground-water observations and analytical results on a regular basis (e.g., quarterly), it is helpful to include a table giving all (or at least the most recent, if voluminous) of the previously collected data for comparison (Table 5-10). It is also helpful to provide changes in such data as the rise or fall of water level, increase or decrease in the thickness of floating product, or the fluctuation in the concentration of a specific dissolved constituent.

REFERENCES

Allison, G. B., W. J. Stone, and M. W. Hughes. 1985. Recharge in karst and dune elements of a semi-arid landscape as indicated by natural isotopes and chloride. *Journal of Hydrology*. 76: 1–25.

Berkstresser, C. F., Jr., and W. A. Mourant. 1966. Ground-water resources and geology of Quay County, New Mexico. Ground-water report 9. New Mexico Bureau of Mines and Mineral Resources. 115 p.

Cruz, R. R., R. K. De Wees, D. E. Funderburg, R. L. Lepp, D. Ortiz, and D. Shaull. 1994. Water resource data, New Mexico, water year 1993. Water-data report NM-93-1. U.S. Geological Survey. 590 p.

Davis, S. N., and R. J. M. DeWeist, 1966. *Hydrogeology*. New York: John Wiley and Sons, Inc. 463 p.

TABLE 5-8 Radiochemical Quality of Ground Water from Wells, Pueblo de San Ildefonso, New Mexico

Station Number and Well Identification	^{3}H ($10^{-6}\mu Ci/mL$)	^{137}Cs ($10^{-9}\mu Ci/mL$)	Total Uranium ($\mu g/L$)	^{238}Pu ($10^{-9}\mu Ci/mL$)	$^{239,240}Pu$ ($10^{-9}\mu Ci/mL$)	Gross Alpha ($10^{-9}\mu Ci/mL$)	Gross Beta ($10^{-9}\mu Ci/mL$)
1 Old Community Well	0.2 (0.2)[a]	31 (59)	44 (0.4)	0.053 (0.019)	0.009 (0.009)	23 (5.)	11 (1.)
3 Pajarito Well (pump 1)	0.1 (0.2)	101 (62)	11 (0.1)	0.016 (0.010)	0.012 (0.007)	4. (2.)	5.8 (0.7)
3 Pajarito Well (pump 2)	0.3 (0.2)	125 (66)	7.6 (0.1)	0.004 (0.004)	0.004 (0.004)	9. (3.)	5.3 (0.7)
8 Halladay Well	0.3 (0.2)	135 (59)	1.4 (0.1)	0.019 (0.019)	0.026 (0.016)	4. (1.)	2.2 (0.4)
9 Eastside Artesian Well	0.4 (0.2)	151 (61)	7.2 (0.1)	0.004 (0.009)	0.004 (0.011)	10. (3.)	2.6 (0.5)
Summary							
Maximum concentration	0.4	151	44	0.053	0.166	23	11
Standard[b]	20[b]	120[c]	32[c]	1.6[c]	1.2[c]	15[b]	50[b]
Maximum as a percentage of standard	2.0	126	138	3.3	13.8	153	22
Limits of detection	0.7	40	1	0.1	0.1	3	3

[a]Counting uncertainties are in parenthesis.
[b]Maximum contaminant level–MCL, used for comparison only (NMEIB 1988, EPA 1989b).
[c]Derived concentration guide applicable to DOE drinking water systems—used for comparison only (see Appendix A).

Source: Hoffman and Lyncoln 1992.

TABLE 5-9 Historical Summary of Ground Water Analytical Results 1989–1994, City of Albuquerque, Fourth Street Yard, 1801 Fourth Street, NW, Albuquerque, New Mexico

Monitor Well	Date Sampled	Benzene (µg/l)	Toluene (µg/l)	Ethylbenzene (µg/l)	Total Xylenes (µg/l)	Total BTEX (µg/l)	MTBE (µg/l)	EDB (µg/l)	EDC (µg/l)	TPH-AS-Gasoline (µg/l)	TCE (µg/l)	1,1-DCE (µg/l)	1, 2-DCE (µg/l)	PERC (µg/l)	1,1 DCA (µg/l)
FSY-1	16-Aug-89	ND	ND	ND	ND	—	NA	NA	NA	NA	14				
	26-Feb-90	ND	ND	ND	ND	—	NA	NA	NA	NA	11				
	17-Dec-90	TRACE	ND	ND	ND	—	NA	NA	NA	NA	12.30				
	16-May-91	ND<0.5	ND<0.5	ND<0.5	ND<0.5	—	ND<1.0	ND<1.0	0.3	NA	NA				
	26-Aug-91	ND	ND	ND	ND	—	ND	ND	ND	NA	11.30				
	7-Apr-93	ND<0.3	ND<0.3	ND<0.3	ND<0.6	—	ND<10	NA	NA	ND<100	NA				
	16-May-94	ND<0.5	ND<0.5	ND<0.5	ND<0.5	—	ND<5	NA	ND<0.5	NA	5	ND<0.5	2	ND<0.5	ND<0.5
FSY-2	16-Aug-89	ND	ND	ND	ND	—	NA	NA	NA	NA	12				
	26-Feb-90	ND	ND	ND	ND	—	NA	NA	NA	NA	11.50				
	17-Dec-90	ND	ND	ND	ND	—	NA	NA	NA	NA	11.60				
	16-May-91	ND<0.5	ND<0.5	ND<0.5	ND<0.5	—	ND<1.0	ND<1.0	ND<0.2	NA	NA				
	26-Aug-91	ND	ND	ND	ND	—	ND	ND	ND	NA	10.70				
	16-May-94	ND<0.5	ND<0.5	ND<0.5	ND<0.5	—	ND<5	NA	ND<0.5	NA	13	ND<0.5	4	ND<0.5	ND<0.5
FSY-3	16-Aug-89	ND	ND	ND	ND	—	NA	NA	NA	NA	30				
	26-Feb-90	17.20	ND	ND	ND	—	NA	NA	NA	NA	49				
	10-Apr-90	25	ND	ND	ND	17.2	NA	NA	NA	NA	44				
	17-Dec-90	7.50	ND	ND	ND	25	NA	NA	NA	NA	50.70				
	16-May-91	ND<0.5	4.6	ND<0.5	ND<0.5	7.5	ND<1.0	ND<1.0	0.2	NA					
	26-Aug-91	500.80	185.30	3.20	57.60	747	ND<5.0	ND	ND	NA	78.40				
	8-Oct-91	488	140.00	ND<100	ND	628	NA	ND	ND	NA	ND				
	6-Mar-92	1,400	76	93	92	1,661	560	NA	NA	NA	NA				
	7-Apr-93	490	11	10	14	530	180	NA	NA	1,600	NA				

NA—Not analyzed
ND—Not detected at indicated detection limit
MTBE—Methyl tertiary butyl ether; BTEX and MTBE per EPA method 602/8020, modified 8020, 5030/8020, or 5030/602
EDB—1,2, - dibromoethane, per EPA method 504
EDC—1,2, - dichloroethane, per EPA method 8010 or 610
TCE—Trichloroethene, per EPA method 8010
1, 1-DCE—1, 1,-dichloroethene per EPA method 8010
1, 2-DCE—1, 2,-dichloroethene per EPA method 8010
1, 1-DCA—1, 1,-dichloroethene per EPA method 8010
PERC—Tetrachloroethene per EPA method 8010
*Laboratory report indicates that hydrocarbons detected in sample in the gasoline range do not match the gasoline standard.

Source: From Hershberger 1994.

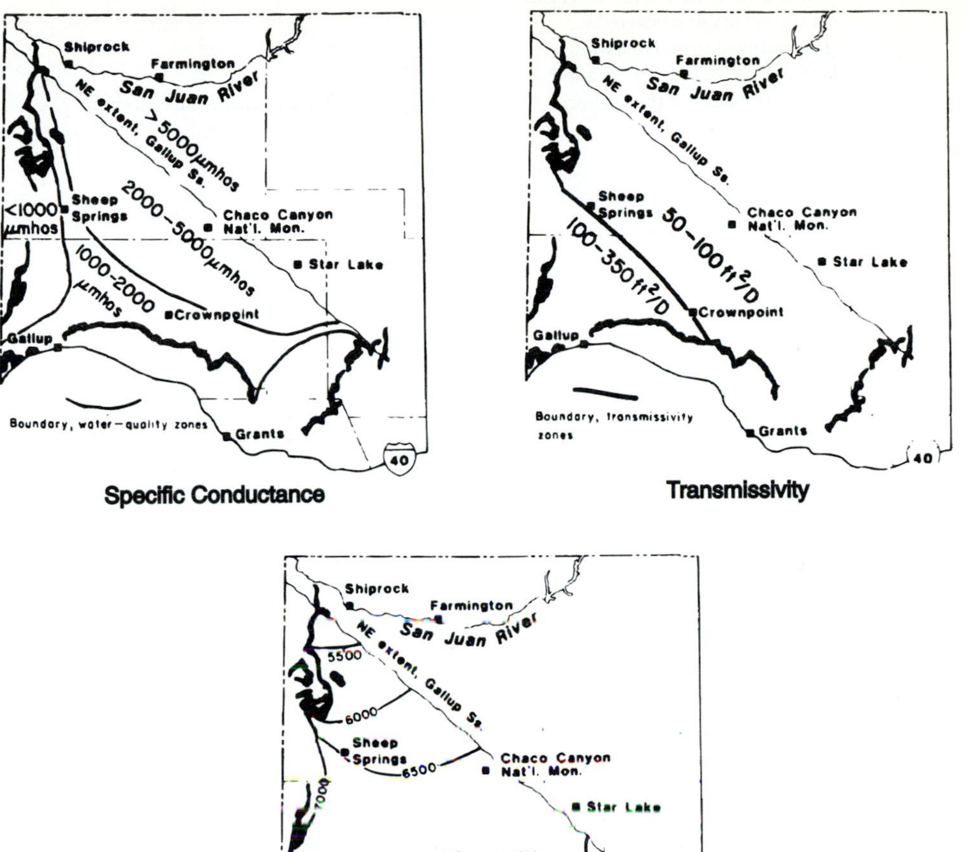

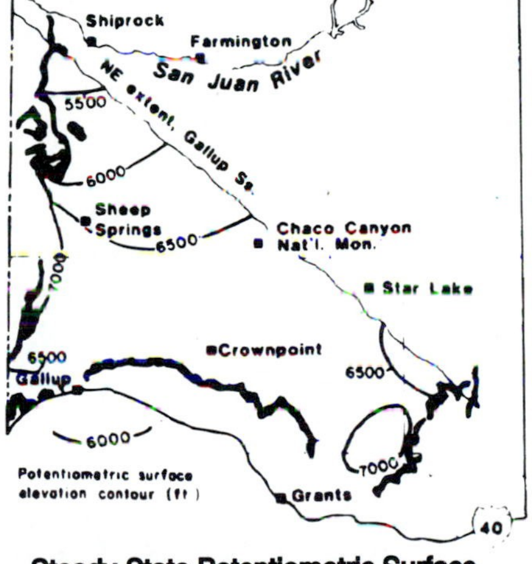

FIGURE 5-6 Suite of maps giving hydrologic conditions for the Gallup Sandstone aquifer in the San Juan Basin, New Mexico. Irregular black areas are outcrops of the Gallup Sandstone. See Figure 2-6 for location. (From Stone 1981, Figure 2. Reprinted by permission of Ground Water Publishing).

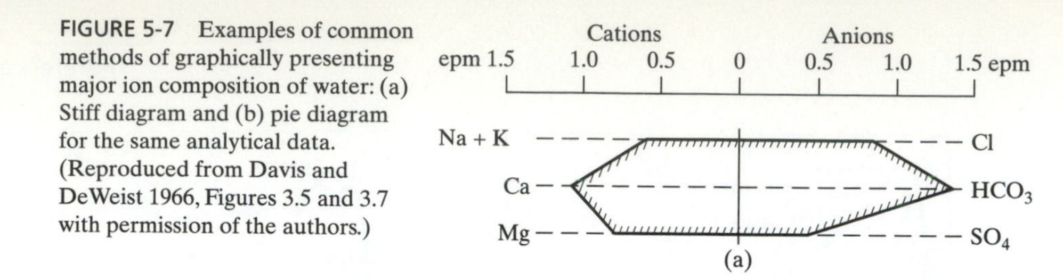

FIGURE 5-7 Examples of common methods of graphically presenting major ion composition of water: (a) Stiff diagram and (b) pie diagram for the same analytical data. (Reproduced from Davis and DeWeist 1966, Figures 3.5 and 3.7 with permission of the authors.)

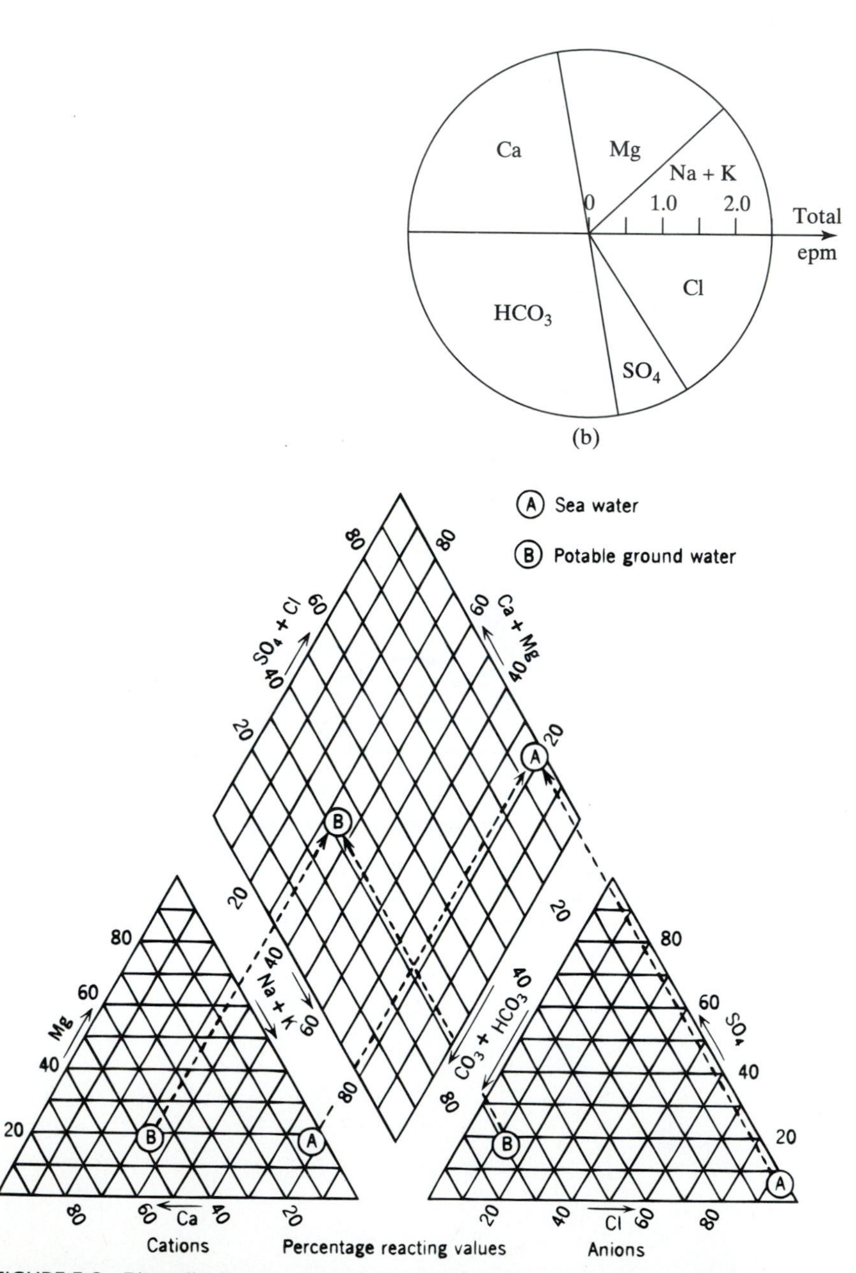

FIGURE 5-8 Piper diagram for classifying and comparing the major-ion content of several waters at once. (Reproduced from Davis and DeWeist 1966, Figure 3.9 with permission of the authors.)

TABLE 5-10 Historical Comparison of Analytical Data from Wells Sampled, AT & SF Railyard, Belen, New Mexico

Well ID	September 1992		March 1993		October 1993		March 1994		September 1994	
	Total PAH (μg/L)	Total BTEX (μg/L)	Total PAH (μg/L)	Total BTEX (μg/L)	Total PAH (μg/L)	Total BTEX (μg/L)	Total PAH (μg/L)	Total BTEX (μg/L)	Total PAH (μg/L)	Total BTEX (μg/L)
MW-5	NA	NA	NA	NA	NA	NA	NA	NA	664	3.8
MW-7	ND	ND	ND	ND	ND	ND	ND	ND	ND	ND
MW-17	210	8.8	93.6	6.5	80.8	7.00	258	5.9	697.6	4.5
MW-24	13.8	12.7	15.34	ND	5.67	ND	15.45	5.0	26.9	1.6*
MW-26	ND	ND	ND	ND	ND	ND	ND	ND	NA	NA
MW-27	ND	ND	NA	NA	NA	NA	NA	NA	NA	NA
MW-39	NA	NA	NA	NA	ND	ND	ND	ND	ND	ND
MW-40	NA	NA	NA	NA	ND	ND	ND	ND	0.07	ND
MW-41	13.1	ND	124.56	ND	NA	NA	NA	NA	NA	NA
MW-45	19.97	ND	0.38	ND	ND	ND	ND	ND	0.91	ND
MW-46	ND	ND	ND	ND	ND	ND	ND	1.0	ND	ND
MW-52	0.16	ND	ND	ND	ND	ND	0.12	ND	ND	ND
MW-53	5.51	ND	ND	ND	ND	ND	ND	ND	0.28	ND
MW-55	6.77	ND	2.12	ND	1.4	ND	0.29	ND	2.32	ND
OW-2	0.69	ND	ND	ND	NA	NA	4.25	ND	ND	ND

*Toluene detected in field blank at approximately same concentration.
NA Not Analyzed.
ND Not Detected.
Source: Radian 1995.

Driscoll, F. G. 1986. *Groundwater and wells.* St. Paul, MN: Johnson Division. 1089 p.

Fetter, C. W. 1994. *Applied hydrogeology.* Upper Saddle River, NJ: Prentice-Hall. 691 p.

Fuller, M. L. 1905. Underground waters of eastern United States. Water-supply paper 114. U.S. Geological Survey.

Harrison, S. S. 1968. The flood problem in Grand Forks/East Grand Forks. miscellaneous series 35. North Dakota Geological Survey. 42 p.

Heath, R. C. 1982. Classification of ground-water systems of the United States: *Ground Water.* 20(4): 393–401.

Heath, R. C. 1984. Ground-water regions of the United States. Water-supply paper 2242. U.S. Geological Survey.

Hershberger, V. 1994. Quarterly monitoring report, September-November 1994. Albuquerque, NM. Consulting report. Groundwater Technology. 9 p.

Hoffman, L., and K. Lyncoln. (compilers). 1992. Environmental surveillance at Los Alamos during 1990. Report LA-12271-MS. Los Alamos National Laboratory, Los Alamos, NM.

Keller, E. A. 1992. Environmental geology. New York: MacMillan. 521 p.

Maulde, H. E., and A. G. Scott. 1977. Observations of contemporary arroyo cutting near Santa Fe, New Mexico, USA. *Earth Surface Processes* 2:39–54.

Meinzer, O. E. 1923. The occurrence of ground water in the United States—with a discussion of principles. Water-supply paper 489. U.S. Geological Survey. 321 p.

Myers, R. G. and R. P. Herman. 1982. Bibliography of dissertations and theses on water and water-related topics from New Mexico Institute of Mining and Technology. Report 156. New Mexico Water Resources Research Institute. 12 p.

The National Academy of Sciences. 1974. More water for arid lands—promising technologies and research opportunities. Report of an ad hoc panel of the Advisory Committee on Technology Innovation, Board on Science and Technology for International Development, Commission on International Relations, The National Academy of Sciences.

Piper, A. M. 1944. A graphic procedure in geochemical interpretation of water analyses. *Transactions of the American Geophysical Union.* 25:914–23.

Purtymun, W. D. 1995. Geologic and hydrologic records of observation wells, test holes, test wells, supply wells, springs and surface water stations in the Los Alamos area. Report LA-12883-MS. Los Alamos National Laboratory. Los Alamos, NM. 339 p.

Radian Corporation. 1995. Remedial monitoring report, third quarter 1995, AT&SF fueling facility and switching yard, Belen, New Mexico. Radian Corporation, consulting report.

Ries, H., and T. L. Watson. 1914. *Engineering geology.* New York: John Wiley and Sons.

Saunders, R. J., and J. J. Warford. 1976. *Village water supply—economics and policy in the developing world.* A World Bank Research Publication. Baltimore, MD: The Johns Hopkins University Press. 279 p.

Stiff, H. A., Jr. 1951. The interpretation of chemical water analyses by means of patterns. *Journal of Petroleum Technology* 3(10):15–16.

Stone, W. J. 1981. Hydrogeology of the Gallup Sandstone, San Juan Basin, northwest New Mexico, *Ground Water* 19(1):4–11.

Stone W. J. 1992. Water-resource information in New Mexico Bureau of Mines and Mineral Resources Reports, New Mexico Bureau of Mines and Mineral Reports booklet. 10 p.

Stone, W. J., T. Leeds, R. C. Tunney, G. A. Cussack, and S. A. Skidmore. 1991. Hydrology of the Carlin Trend, northeastern Nevada—a preliminary report: company report, Hydrology Department, Newmant Gold, CO. 123 p.

Stone, W. J., F. P. Lyford, P. F. Frenzel, N. H. Mizell, and E. T. Padgett. 1983. Hydrogeology of San Juan Basin, New Mexico. Hydrologic report 6. New Mexico Bureau of Mines and Mineral Resources. 70 p.

Thomas, H. E. 1954. Ground-water regions of the United States—their storage facilities. *The physical and economic foundation of natural resources*, vol. 3. U.S. Congress, House Committee on Interior and Insular Affairs.

Todd, D. K. 1980. *Groundwater hydrology*. New York: John Wiley and Sons. 535 p.

Walker, G. R., I. D. Jolly, M. H. Stadter, F. W. Leaney, W. J. Stone, P. G. Cook, R. F. Davie, and L. K. Fifield. 1990. Estimation of diffuse recharge in the Naracoorte Ranges region, South Australia—an evaluation of chlorine-36 for recharge studies. Report of AWRAC Research Project P87/10, Australian Water Research Advisory Council. 73 p.

Walton, W. C. 1970. *Groundwater resource evaluation*. New York: McGraw-Hill. 664 p.

CHAPTER 6

Characterizing Hydrologic Conditions

Once you have compiled hydrologic information, it must be integrated into a clear characterization of the regional water system. This characterization includes all parts of the hydrologic cycle. The ultimate goal of an hydrogeologic study is an understanding of the interaction of the surface-water, soil-water, and ground-water components. This understanding is communicated through a sound description of the hydrologic setting. The key to adequately characterizing the hydrologic conditions lies not so much in getting the right answers, as in asking the right questions.

HYDROLOGIC PHENOMENA

We all know that the amount of water on earth is essentially fixed. Water's tendency to constantly change form and location is referred to as the hydrologic cycle (Figure 6-1). The word *cycle* is not meant to imply that a given molecule of water makes the entire journey from ocean to land and back to the ocean uninterrupted. In reality, the cycle may be short-circuited at any time, especially whenever water is returned to the atmosphere. Rather, *cycle* refers to the reprocessing of a fixed volume of water on the planet.

The cooling of clouds laden with oceanic water vapor moving across the land produces fog, dew, rain, sleet, hail, or snow, depending on the temperature and other factors. The water discharged in these various forms of precipitation is redistributed by several processes. After it snows, some water may be returned directly to the atmosphere through sublimation; that is, water may pass from solid to gas, without going through the liquid phase. When water comes from rain and melting snow or ice, some of it moves along the surface by overland or channelized flow (runoff) to rivers. Sometimes these rivers may deliver the water to standing-water bodies, where evaporation may return the water to the atmosphere. Otherwise, rivers eventually return the water to the sea to start the process all over again. Some of the water from snow and rain soaks into the soil, sediment, or rock at the surface or beneath streams and lakes (infiltration). Much is returned directly to the atmosphere as vapor from the geologic materials at the surface and from rivers and standing-water bodies (evaporation) or indirectly from plants (transpiration). Infiltrated water that escapes evapotranspiration moves downward (percolation) until it is added to the region's ground water (recharge). Flow in the saturated zone may return the ground water to a stream, a swamp, a lake, or even the ocean.

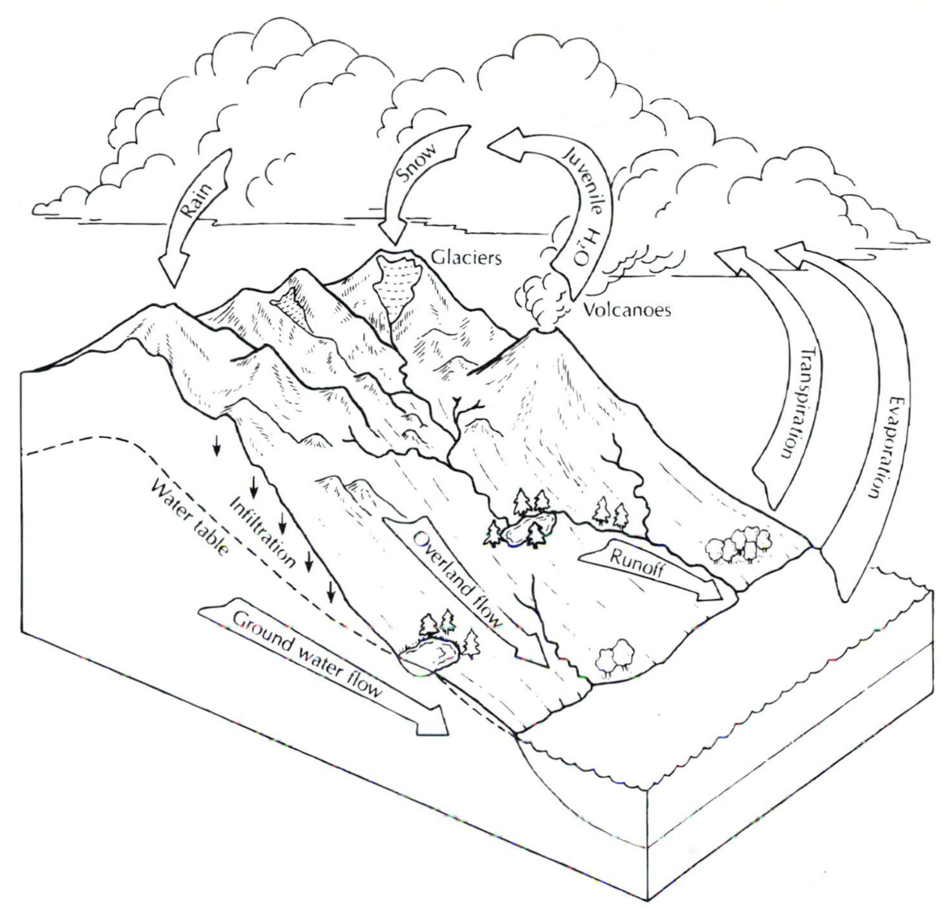

FIGURE 6-1 The hydrologic cycle. (From Fetter 1994, Figure 1.4.)

SURFACE WATER

It is important to understand the surface-water system of the study area, even if the focus seems to be on ground water. This system, including flowing and standing surface water, makes up a large part of the hydrologic cycle. It interacts with soil and ground water and gives rise to unique environmental conditions. For example, the surface-water system disperses runoff, makes riparian habitat, supports aquatic wildlife, and provides for water-based recreation. Surface water is also a major pathway for contaminant transport.

A stream may gain or lose water along its course, depending on the depth of the water table (Figure 6-2). Where the top of the saturated zone is very shallow, as in humid regions, water may be gained through ground-water discharge. Where the water table is deep, as in arid regions, water may be lost to the soil-water and ultimately, perhaps, to the ground-water systems.

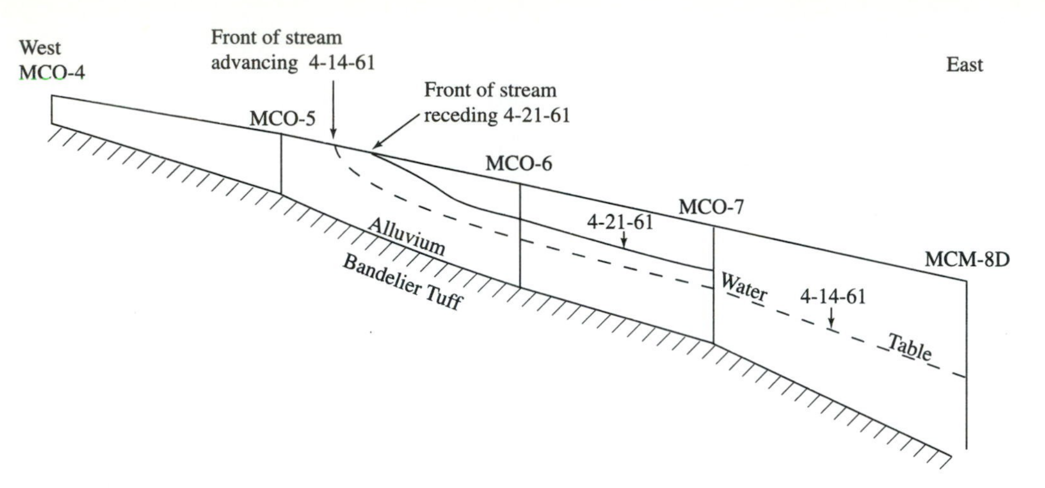

FIGURE 6-2 Relationship of surface water to shallow perched ground water in a canyon cut into the Pajarito Plateau, north-central New Mexico (as modified from Baltz and others 1963, Figure 8; by Stone 1995, Figure 5).

Flowing Water

When characterizing the surface-water system, a good place to start is with a description of the watershed in which the study area lies and the streams within it. In this description, answer several basic questions.

What is the size of the area drained?

What stream drains the area directly?

If not a main stream, of what larger system is the local stream a part?

What type of drainage pattern prevails?

Is the channel meandering or braided?

How wide is the channel (average, range)?

What are the predominant channel and bank materials?

In addition to providing written descriptions of these characteristics, it is a good idea to include illustrations as well. The types of tables and figures discussed in "Useful Hydrologic Tools," Chapter 5, may also be applied here.

Next, describe the main stream's hydraulic characteristics.

What type of flow is involved (perennial or ephemeral; continuous or intermittent)?

Where is it gaining and where is it losing (Figure 6-3)?

What is known of stage versus discharge (Figure 6-4)?

What are the maximum, minimum, and average discharges?

How were these measured?

How many years of record are there?

Supplement this description with appropriate tables and figures.

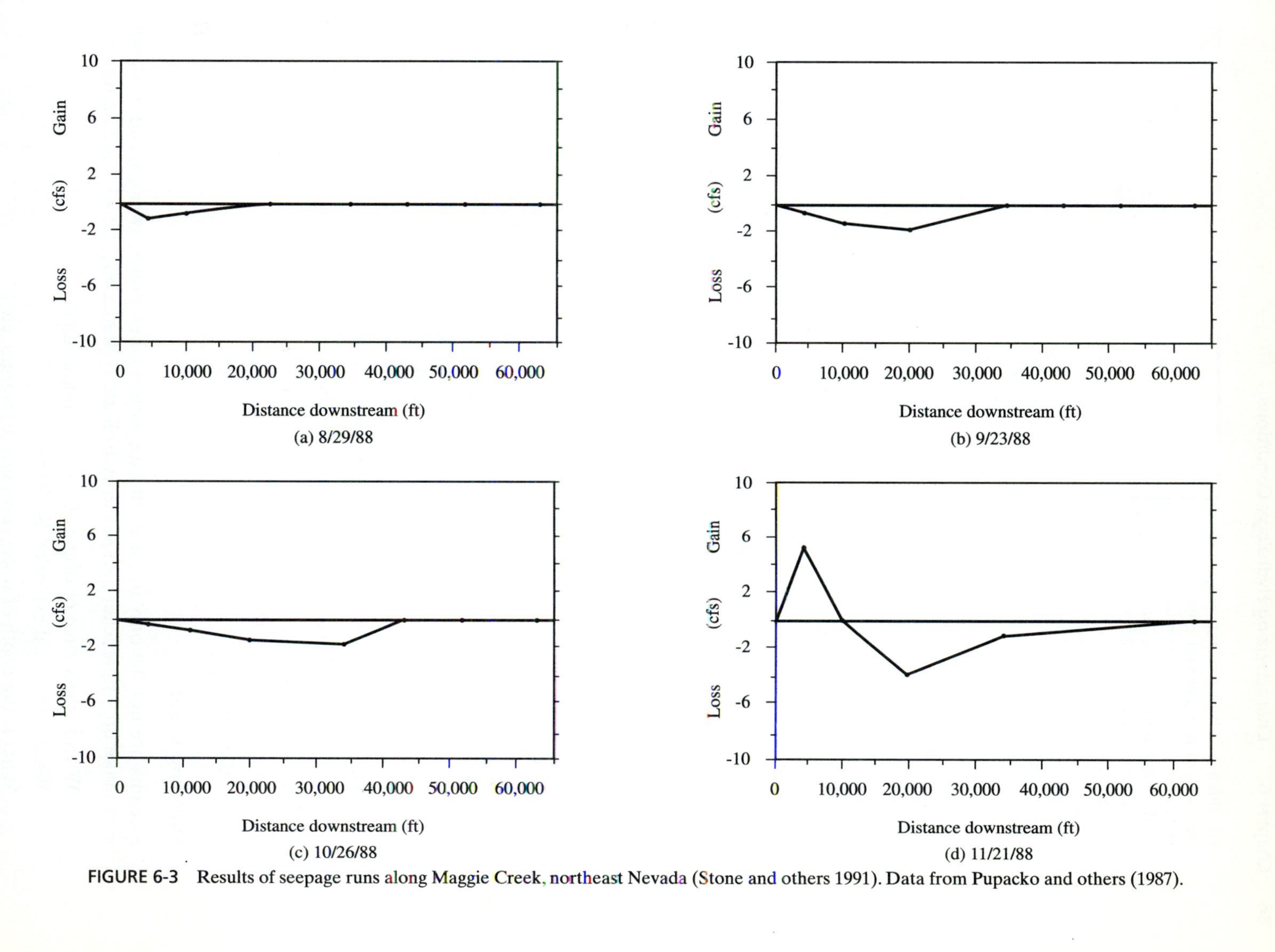

FIGURE 6-3 Results of seepage runs along Maggie Creek, northeast Nevada (Stone and others 1991). Data from Pupacko and others (1987).

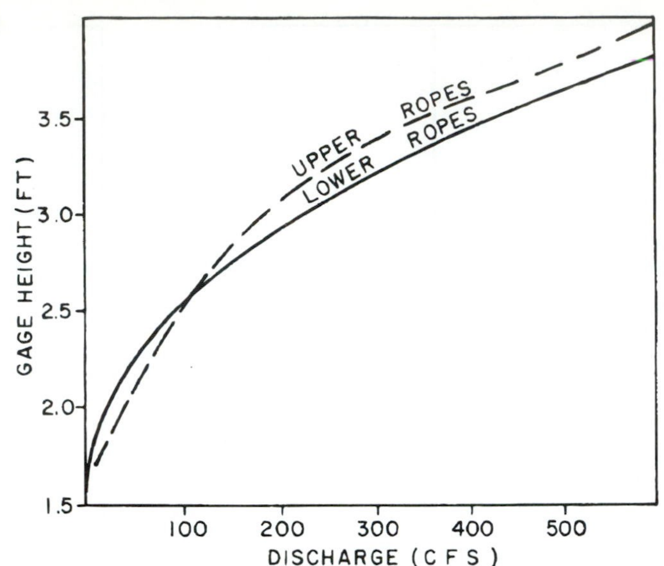

FIGURE 6-4 Stage-discharge rating curves for stream gages on Ropes Draw, a small semiarid watershed on the western flank of the San Andres Mountains, south-central New Mexico. (From Stone and Brown 1975, Figure 7. Reproduced with permission of the New Mexico Geological Society.)

Flow frequency, or the percent of the time that a given discharge will be met or exceeded, should also be addressed. The higher the discharge, the lower the frequency or less likely it is to occur.

This is most easily illustrated with a flow-frequency curve (Figure 6-5). To ensure that readers understand such graphs, in the text give examples of discharges expected for selected frequencies based on them. Summarize the flow frequency of various streams in the area and compare them (Table 6-1).

Finally, include indications of the water's chemical characteristics.

What is known about the general water-quality parameters, such as temperature, pH, suspended load, bedload, specific conductance, total-dissolved-solids content, dissolved oxygen content, and biological oxygen demand?

What is the concentration of major ions, metals, radionuclides, and organic constituents?

How do these parameters or concentrations vary with the season?

How have they varied through the period of record?

Tables may be used to list specific data, while figures are useful for showing their variation with time. In spite of such illustrations, the text should give ranges and possibly mean values for water-quality indicators.

Standing Water

Streams are not the only forms of surface water; lakes, ponds, swamps, and wetlands may make up a significant part of the hydrologic system. Describe all standing water.

What is the source or origin of the standing water?

What is the areal extent of the feature?

What is the depth of water (average, maximum, etc.)?

What is known about the water quality (as for streams)?

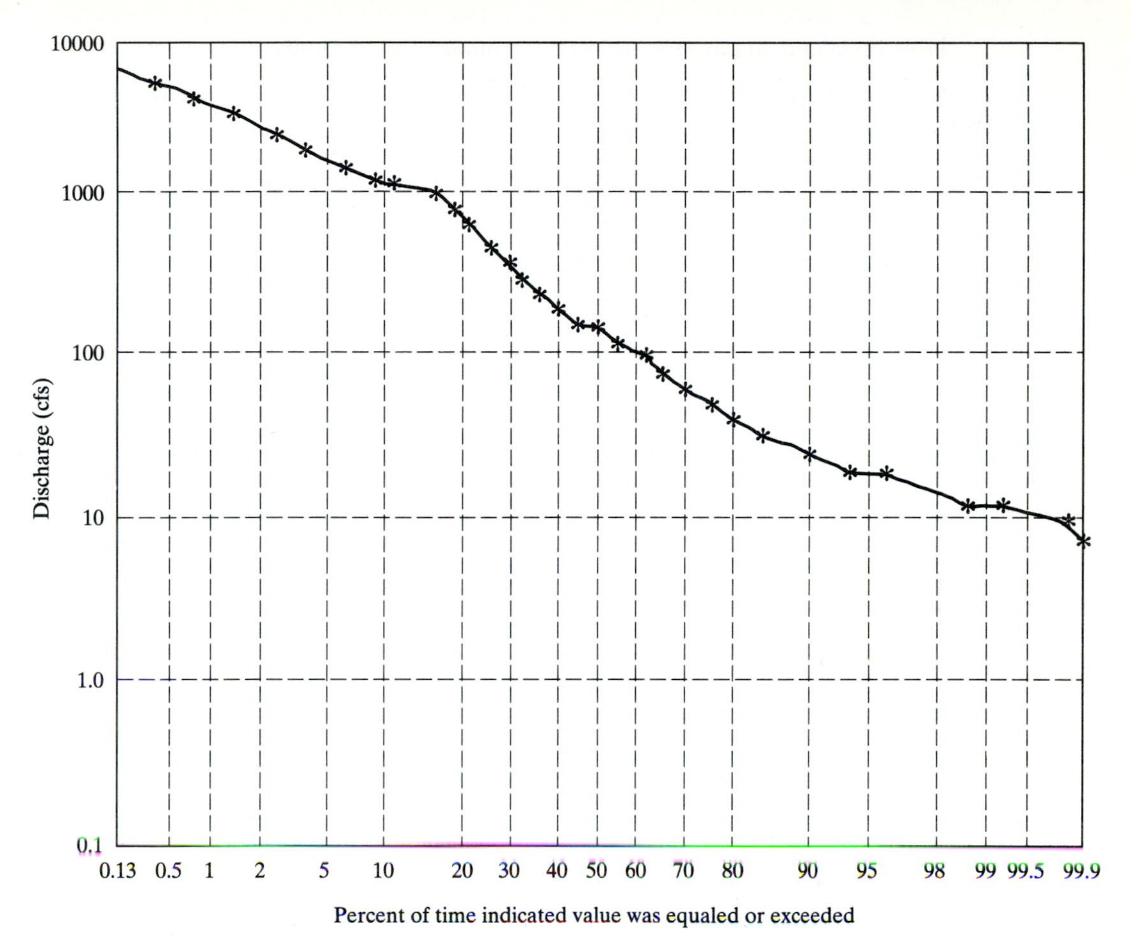

FIGURE 6-5 Flow-frequency curve for the Humboldt River at Palisades, northeast Nevada, 1940–1989. (From Stone and others 1991, Figure 8; data from USGS.)

TABLE 6-1 Comparison of Flow-Frequency Values for Area Streams, Carlin Trend, Nevada

	Portion of Time (%) Indicated Discharge is Equaled or Exceeded				Discharge (cfs) Expected Indicated Portion of Time						
Stream	1 cfs	10 cfs	100 cfs	1,000 cfs	1%	5%	10%	25%	50%	75%	95%
Humboldt River at Carlin	99.4	92	55	11	5,000	3,000	1,500	700	200	50	9
Maggie Creek at Carlin	72	30	8	0	>190	140	80	13	3.5	1.5	<1.0
Humboldt River at Palisade	100	99.7	62	17	5,500	3,100	1,700	750	300	75	30
Rock Creek at Battle Mountain	78	38	10	0.5	800	250	100	40	7.5	1.5	<0.1

Source: From Stone and others, 1991.

How do chemical parameters vary seasonally and annually?

Some of the same types of illustrations used for flowing water may be useful for standing water.

SOIL WATER

Water in the unsaturated or vadose zone has traditionally been called *soil water,* regardless of whether or not the geologic material is actually soil. It is replenished by direct infiltration of precipitation as well as seepage from surface runoff and, in some cases, standing water bodies. Once within the vadose zone, soil water may be taken up by plants (transpiration), may be returned to the atmosphere (evaporation), or may percolate beyond the root zone and reach the water table (ground-water recharge).

The soil-water system is important because it supports natural and cultivated plant life. Also, in the case of leaks or spills, it will be contaminated before the saturated zone. An understanding of the unsaturated zone in a study area, as well as monitoring and early detection of such problems, may permit their remediation before ground water is impacted.

Go over the physical or chemical parameters needed to characterize the soil water in the study area by asking:

What is the moisture content (amount of water present)?

What is the hydraulic conductivity of the soil (unsaturated and saturated)?

What is the tension (matric suction)?

What is the chemistry of the soil water (amount of various dissolved species, such as nitrate)? As we will discuss next, the chloride content of the soil water can be used to determine ground-water recharge.

Include these parameters when the information is available, especially if the unsaturated zone is critical to the problem under study.

GROUND WATER

Most water-resource and environmental investigations deal mainly with ground water. Water-availability and contaminated-site studies must thoroughly describe the ground-water system. This task may be broken down into three parts: describing ground-water occurrence, movement, and quality.

Occurrence

One of the first things to determine and report is the location of the ground water. Its location may be described both geologically and geographically. Described geologically, significant ground water in an area occurs in one or more geologic units (aquifers). Geographically, water may be more available in certain areas owing to variations in geology or topography.

The approach used to identify aquifers in an area where wells already exist differs from that used in an area where there has been little or no previous drilling. In areas where there are nearby wells, a review of existing reports or files should help identify the aquifer(s) involved. Refer to cross sections as well as structure, depth, and thickness maps for the area aquifers. Lacking these, the aquifer may be identified using a geologic map and stratigraphic column of the area. Based on the unit at the surface (from the geologic map) and the observed depth to water, the aquifer may be identified from the thickness of units as shown on the rock column.

Movement

Next, describe ground-water movement. This description consists of three components: recharge, flow, and discharge.

For recharge, determine and report the location of the recharge area, the recharge mechanism, and the recharge rate. Recharge areas are those in which water movement is downward, that is, away from the ground surface and toward the water table. Location will be suggested by the water-level map: the recharge area will be upgradient. When a water-level map is not available, use a topographic map for general purposes: recharge occurs in high elevations. If drilling is an option, use two closely spaced wells of different depths (paired wells). In recharge areas, the water level in the deeper well will be lower than that in the shallower well, indicating a downward vertical gradient.

The recharge mechanism may be deduced by examining geomorphic and structural features in the area. For example, if there are any sinkholes, playas, faults or joints, recharge may be localized. Recharge is also brought about by the seepage of water along stream beds (transmission loss), provided plants do not capture the moisture before it moves below the root zone. If such direct pathways are not apparent, recharge must be more regional and consist of the areawide take-up of precipitation. Although recharge begins with infiltration, they are not the same. Water added to the surface by infiltration may be redistributed by evapotranspiration. Only deeper percolating water can recharge the ground water.

The rate of recharge may be determined by various methods (Allison 1988). Mainly, these methods fall into two categories: physical methods and chemical methods. An example of a physical method is measuring the advance of a soil-moisture wetting front with a neutron probe. Chemical methods rely on various natural and introduced tracers. Chloride is an example of a natural tracer that has been used for determining modern as well as paleorecharge rates (Anderson 1945; Allison and Hughes 1978; Stone 1992). Products of atmospheric atomic-bomb testing are indirectly introduced tracers. For example, the distribution of tritium and chlorine-36 with depth in the vadose zone has been used to determine downward movement of soil water or recharge rate (Phillips 1994). Directly introduced tracers include various dyes, salts (bromide), and radioactive substances. The stable isotopes ^{18}O and deuterium (^{2}H) are often used in conjunction with natural or introduced tracers to further interpret the movement of water in soil (Clark and Fritz 1997).

Ground-water flow has both a direction and a rate. Direction is best determined from a water-level map. Generally, water flows from high water level to low water level and perpendicular to water-level contours (Figure 6-6). Take care to make separate water-level maps for different aquifers, especially for shallow unconfined and deeper confined systems (Figure 6-7). Figure 6-7a illustrates the consequences of not doing this.

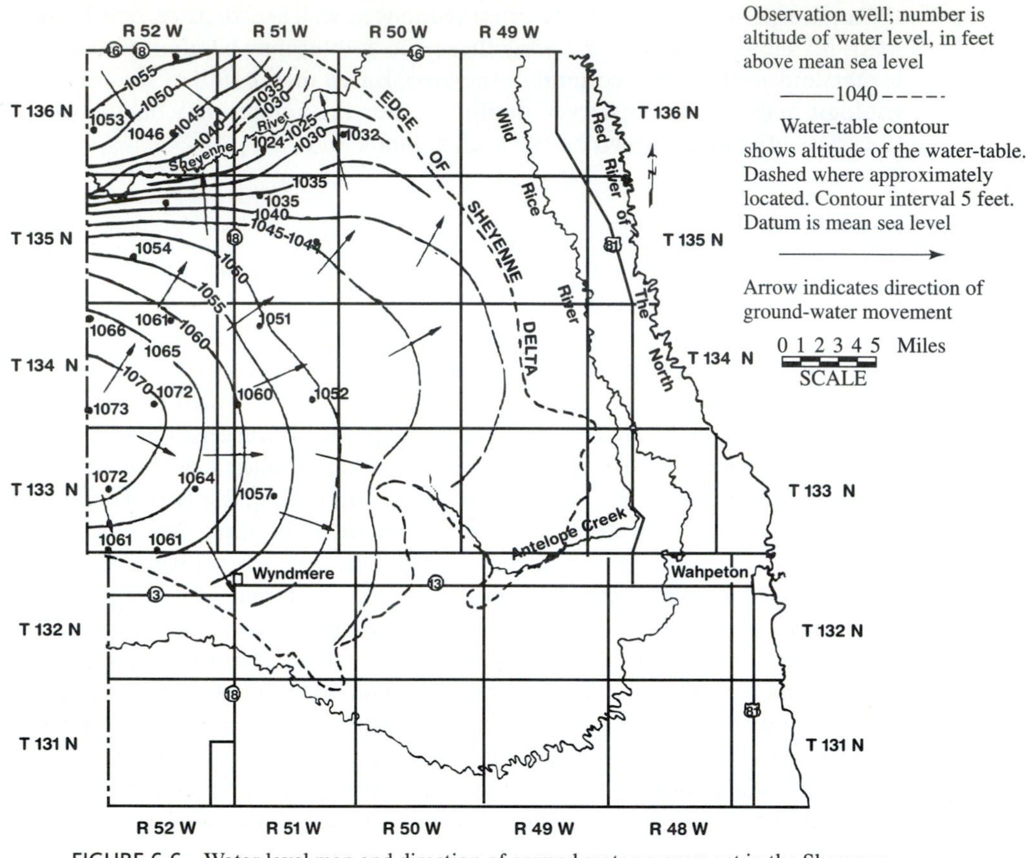

FIGURE 6-6 Water-level map and direction of ground-water movement in the Sheyenne Delta region of southeastern North Dakota. (From Baker and others 1967, Figure 8.)

If multiple aquifers are suspected, plot the water-level elevations against distance along a suspected flow line (a transect perpendicular to water-level contours), using a different symbol for each aquifer. If the aquifers are hydraulically separated, it will be apparent from such water-level profiles (Figure 6-8).

Ground-water flow rate (Q) depends on hydraulic conductivity and gradient, according to Darcy's law:

$$Q = -KIA$$

where K is the hydraulic conductivity (negative sign simply indicates that flow of water is in the direction of decreasing head), I is the hydraulic gradient, and A is the cross-sectional area, perpendicular to flow direction. The hydraulic conductivity of an aquifer is its capacity to transmit water. This capacity is usually given in terms of distance per time (for example, ft/day), like a velocity. The actual rate of movement of water in an aquifer, however, depends on the hydraulic gradient or change in water level or head with change in distance. Gradient is thus given in units of length per length (for example, ft/mile).

As noted by Freeze and Cherry (1979), ground-water velocity may be viewed at different scales: the macroscopic or average flow, in contrast to the microscopic velocity of individual water particles along their actual flow paths in pore spaces. Two common velocities used in ground-water hydrology, specific discharge and average linear velocity, are macroscopic.

Specific discharge (Darcy velocity, Darcy flux) is derived from Darcy's Law as the hydraulic conductivity times the gradient,

$$v = KI \quad \text{or}$$

the flow rate divided by the cross-sectional area,

$$v = \frac{Q}{A}$$

However, flow only occurs in the void portion of the cross-sectional area, thus the average linear velocity (seepage velocity) is the flow rate divided by the porosity (n) times the cross-sectional area,

$$v_1 = \frac{Q}{nA}$$

This is the same as specific discharge divided by porosity.

There is a tendency lately to substitute the word *gradient* for *flow direction*. This suggests some misunderstanding of gradient. Gradient is a vector and thus has both magnitude and direction. More specifically, it is the rate of change in water level with change in lateral position in the direction of ground-water flow. Although it is associated with and calculated along the flow direction, it is not, in itself, that direction. Flow direction is deduced from a water-level map. Thus, the proper response to the question, What is the gradient here? is a value, not a compass heading.

Gradient may be reported in either of two ways, with or without units of measure. Units of measure must be included if those used for vertical and horizontal distances are different: 369 ft/mile. However, units of measure are not required if those used for both distances are the same, for example, 0.07. If water level drops 369 ft over a distance of a mile (5,280 ft), dividing the vertical-distance term (in ft) by the horizontal-distance term (in ft) yields a value of 0.07 ft/ft. Since the units are the same, they are not necessary. This value also applies to mi/mi, m/m, or park benches/park bench. Such values are especially useful for comparing gradients in different areas.

Like recharge, discharge refers to an area, a process, and a rate. Discharge areas are places where ground-water flow is largely upward toward the water table or ground surface. Such areas will be associated with the lowest value contours on a water-level map or the lowest elevations on a topographic map. Typically, they result in springs, streams, or wetlands. In arid lands, discharge areas are marked by isolated stands of lush vegetation in an otherwise barren landscape. Paired wells, in this case, would reveal an upward vertical gradient (the water level in the deeper well would be higher than that in the shallower well). The location of discharge areas may be determined from hydrogeologic cross sections (oriented parallel to ground-water flow direction) showing the water table or potentiometric surface (in the case of confined aquifers). If the discharge

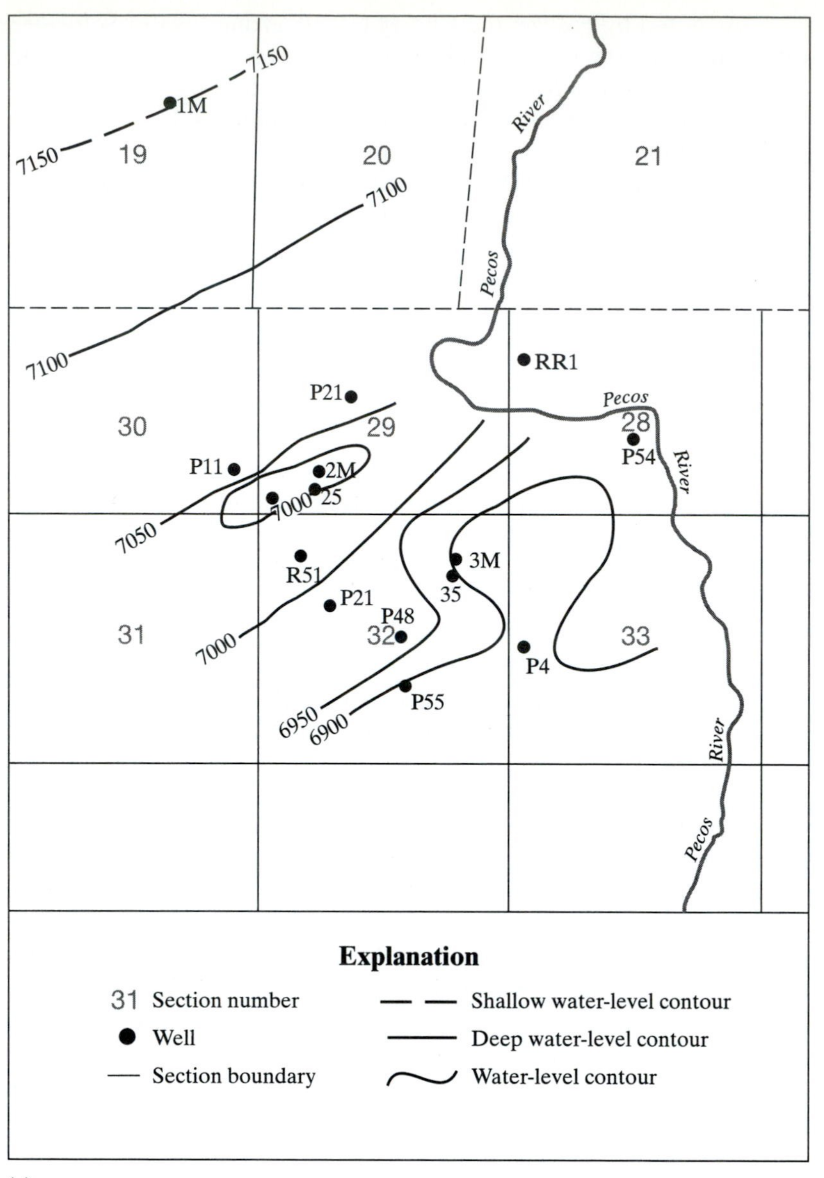

(a)

FIGURE 6-7 Water-level maps for an area of multiple aquifers near Pecos, New Mexico; (a) overly complex map resulting when water levels for two different aquifers (Table 6.2) are contoured together and (b) more reasonable map resulting when water levels for the two aquifers are contoured separately. (Data from the files of New Mexico Environment Department.)

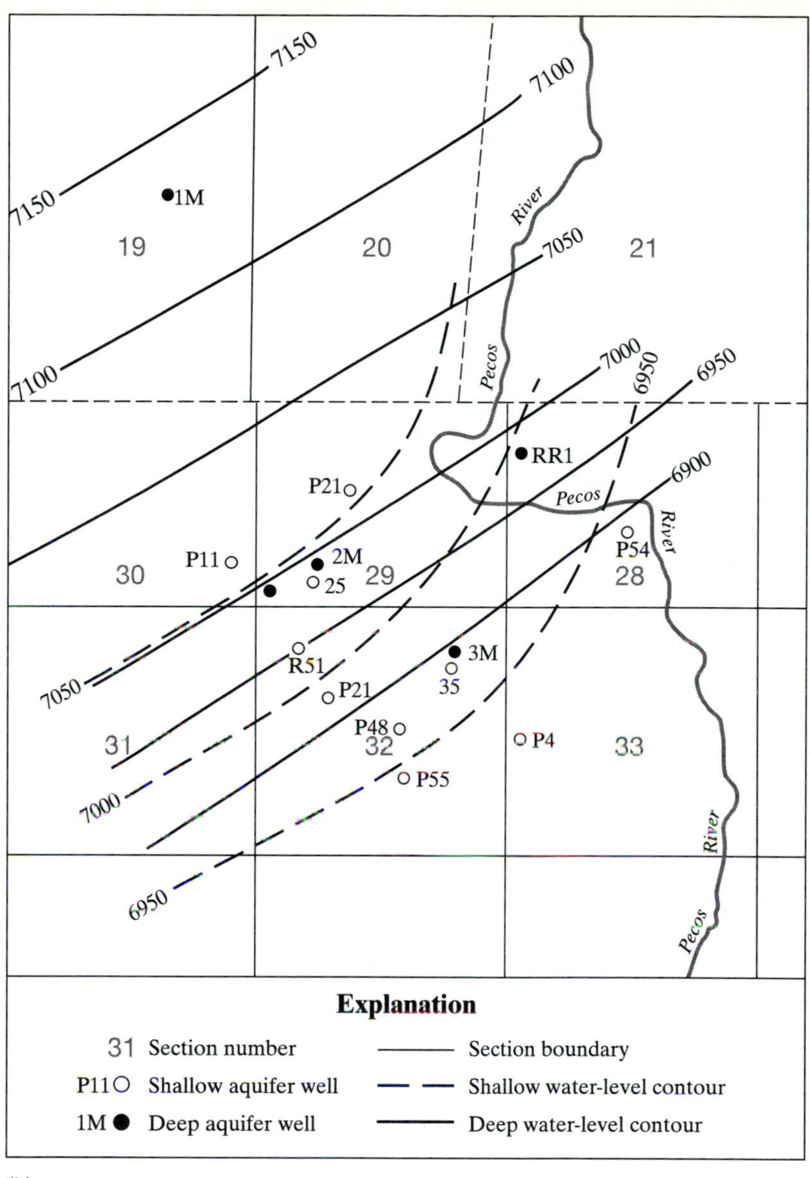
7150
7150
7100
7100
7050
7050
7000
7000
6950
6950
6950
6900
1M
RR1
2M
3M
P21
P11
25
R51
P21
35
P48
P55
P4
P54
19
20
21
30
29
28
31
32
33
Pecos River
Pecos
Pecos
River
Pecos
River
Explanation
31 Section number
P11 Shallow aquifer well
1M Deep aquifer well
Section boundary
Shallow water-level contour
Deep water-level contour

(b)

TABLE 6-2 Water Levels in Wells Near Pecos, New Mexico

Well	Aquifer	Water-Level Elevation (ft)	Data
1M	Deep	7146	10/87
2M	Deep	6972	10/87
2S	Shallow	7039	10/87
3M	Deep	6896	10/87
3S	Shallow	6961	10/87
P4	Shallow	6883	8/85
P11	Shallow	7064	8/85
P14	Deep	6998	8/85
P21	Shallow	6923	8/85
P27	Shallow	7080	8/85
P48	Shallow	6967	8/85
P54	Shallow	6924	8/85
P55	Shallow	<6860	8/87
RR1	Deep	6978	8/87
RS1	Shallow	7012	8/87

Source: Files of the New Mexico Environment Department.

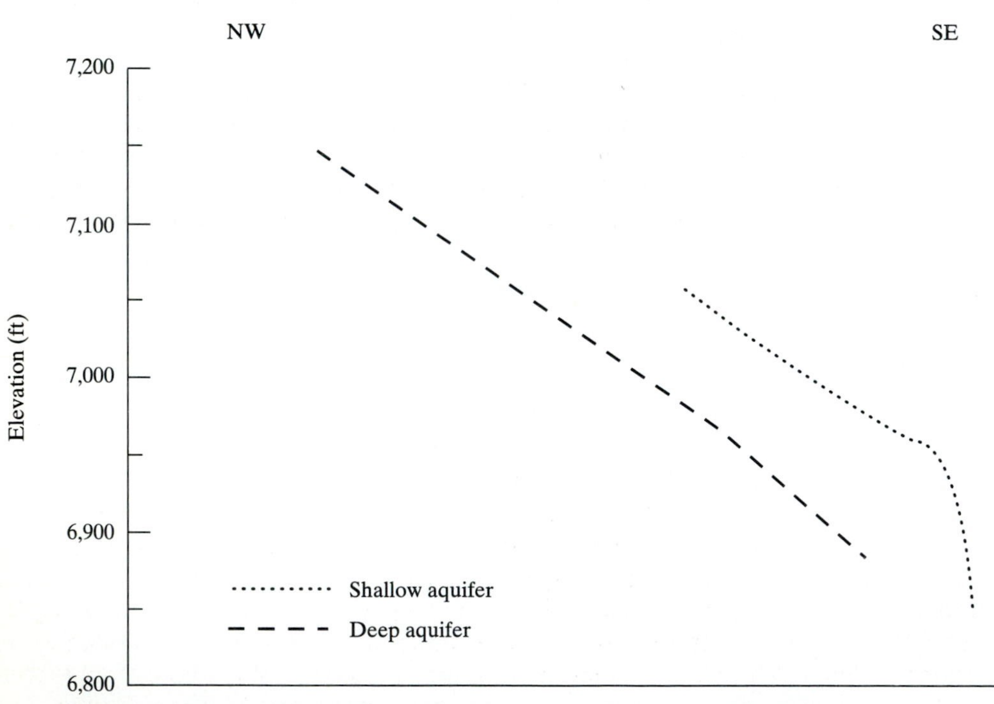

FIGURE 6-8 Water-level profiles constructed for the wells in Figure 6-7. Clearly the two aquifers are hydraulically separated.

area does not lie within the area depicted by the cross section, the slope of the water table or potentiometric surface will at least suggest its general location.

Discharge mechanisms may be natural or artificial. Springs are natural ground-water discharge features. Wells and well fields are artificial means by which ground-water discharge occurs.

Springs occur wherever the ground surface intersects a perched or regional water table or a conduit to a confined water body. The discharge at springs is generally under the influence of gravity, but artesian pressure may be involved in some cases. Various geologic and topographic conditions cause springs, but all fall into one of three general categories:

1. depression springs—the water table merely intersects the ground surface (regardless of whether the aquifer has intergranular or fracture porosity);
2. contact springs—ground water (often perched) issues from a contact where permeable material overlies or abuts impermeable material (the contact may be the result of deposition, faulting or igneous intrusion); or
3. artesian springs—confined water reaches the surface via fractures or faults.

Several types of springs may occur in the same area (Figure 6-9). Describe the various conditions responsible for the springs in a study area.

Water is artificially discharged by pumping, thus well fields should also be considered as discharge areas. Include the location of such pumping areas in the discussion of ground-water discharge.

Discharge rate may be determined in various ways, depending on the processes involved. For example, a spring's rate of flow may be measured with a flume or weir. Ground-water discharge to gaining reaches of streams (those along which flow increases) may be determined from seepage runs (see Figure 6-3). The rate of artificial discharge associated with pumping may be determined by metering well output or estimated from interviews with well owners. In a regional sense, natural discharge equals recharge; when one is known, the other is known.

Quality

Finally, discuss the chemical character of the ground water. This discussion should include general parameters as well as the specific constituents present and their concentration. Describe methods of sampling and analysis.

General chemistry includes such qualities as pH, temperature, specific conductance, total-dissolved-solids content (TDS), etc. Results of comprehensive water analyses are usually reported according to four categories: major ions, trace metals, radionuclides and organic compounds. Clearly state whether the results are for field or laboratory measurements and whether they are total (unfiltered sample) or dissolved (filtered samples) values. Parameters of particular interest may be mapped (Figure 6-10). Mention the water's unsuitability for certain purposes and any specific standards exceeded.

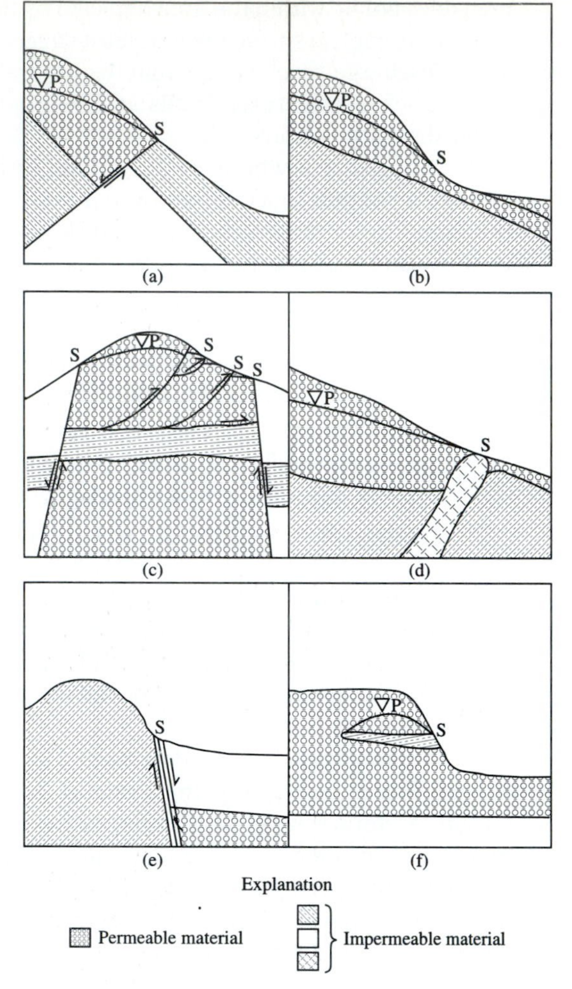

FIGURE 6-9 Schematic diagrams of spring types in the Carlin Gold Trend, Nevada. (P = perched water level, S = spring or seep). (a) fault contact, (b) depression, (c) fault/shallow artesian, (d) intrusion contact, (e) fault/deep artesian, (e) stratigraphic contact. (From Stone and others 1991, Figure 23.)

WATER BALANCE

Once the major components of the hydrologic cycle in the study area have been characterized, it remains only for you to quantify them. The result is called the *water balance* or *water budget* for the area. More specifically, assign annual values to the basic input/output equation

$$P = RO + ET + R$$

where P is precipitation, RO is runoff, ET is evapotranspiration, and R is recharge.

You may obtain data required for this from various sources. Precipitation values may be obtained from the state climatologist or measured on site. Streamflow data collected by a government agency (the USGS in the United States) or measured and calculated on site may be used for runoff. USGS surface-water observations are published

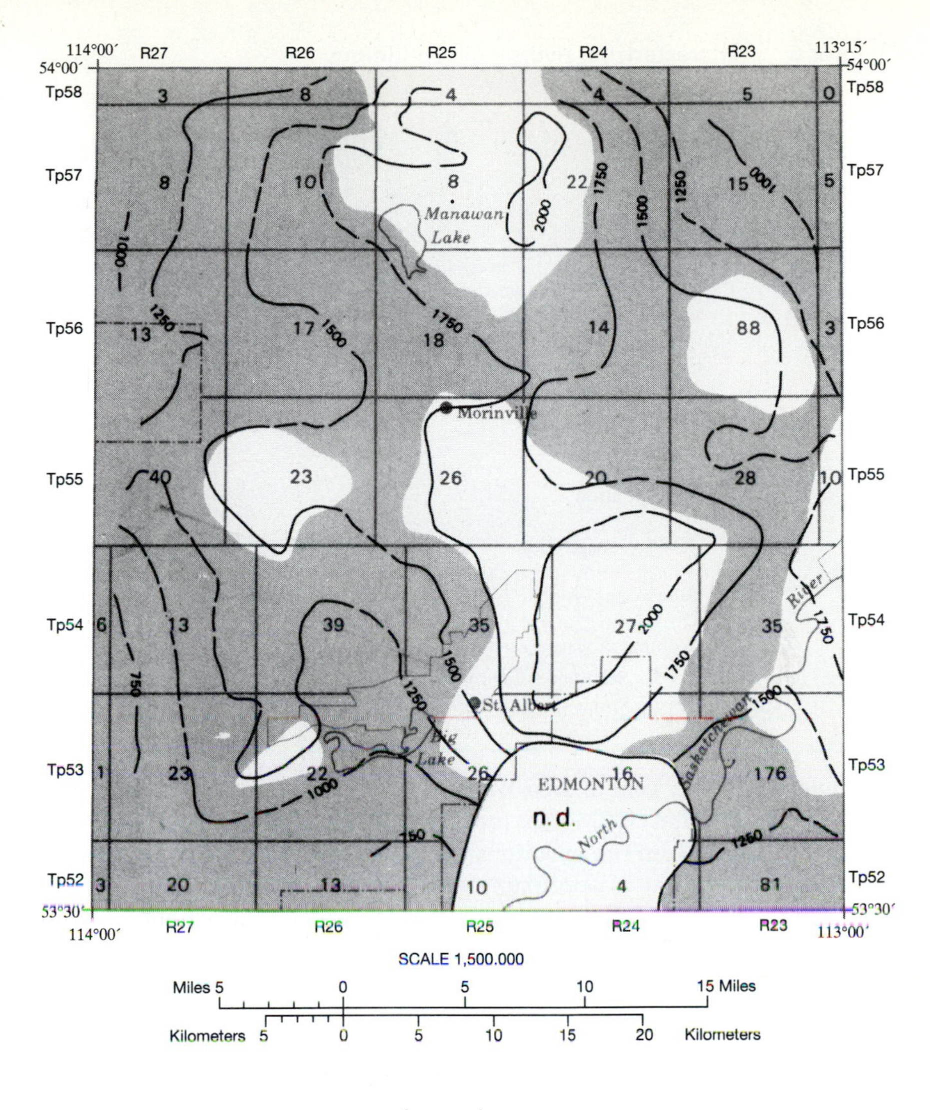

Legend

Number of data points used in each township and range, 3

Total dissolved solids in parts per million

defined, —1000—

approximate, – –1000– –

Carbonate + bicarbonate constituting over 60% of total anions,*

Area of no data n.d.

Over entire map area calcium + magnesium constitute less than 60% of total cations* and sodium + potassium constitute more than 60 % of total cations*

*Determined on equivalents per million basis.

FIGURE 6-10 Hydrochemical map for Edmonton, Alberta, Canada, emphasizing the major anion species and total-dissolved-solids content of bedrock ground water. (From Bibby 1974, unnumbered figure on hydrogeologic map sheet.)

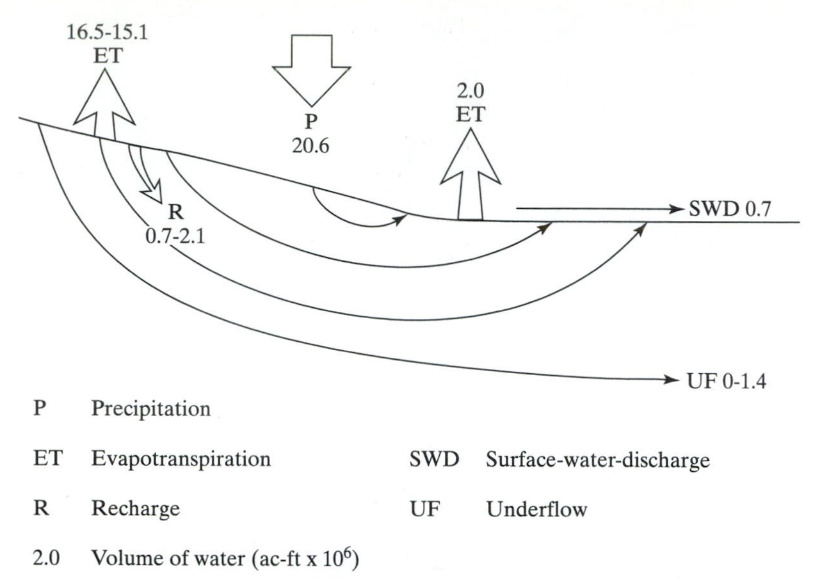

FIGURE 6-11 Schematic diagram of water budget for the Rio Grande Valley in New Mexico. (From Stone and Summers 1987, Figure 4.)

annually by water year in their *Water Resources Data* series. You may also use published evapotranspiration (ET) values for similar vegetation in other regions. Alternatively, you may estimate ET on site in various ways (Shuttleworth 1993). The eddy-correlation and Bowen-ratio methods approximate actual evapotranspiration. The eddy-correlation method uses atmospheric fluxes, whereas the Bowen-ratio method depends on temperature and humidity gradients. The Penman method uses aerodynamic principles to estimate potential ET (Van Hylckama 1980). All these methods focus on the energy balance. Likewise, you may assign a recharge rate from published values for similar areas or calculate it on site by the physical or chemical means described previously. The resulting water budget may be presented in a diagram to make it easier to visualize (Figure 6-11).

REFERENCES

Allison, G. B. 1988. A review of some of the physical, chemical, and isotopic techniques available for estimating groundwater recharge. In Estimation of natural groundwater recharge. NATO ASI Series © *Mathematical and Physical Sciences* 222:49–72.

Allison, G. B., and M. W. Hughes. 1978. The use of environmental chloride and tritium to estimate total recharge to an unconfined aquifer. *Australian Journal of Soil Research* 16:181–195.

Anderson, V. G. 1945. Some effects of atmospheric evaporation and transpiration in the composition of natural waters in Australia 4—underground water in riverless areas. *Journal of Australian Chemical Institute* 12:83–98.

Baker, C. H., Jr. and Q. F. Paulson. 1967. Geology and ground water resources, Richland County North Dakota—Part III. *Ground Water Resources*. Bulletin 46, Part III. North Dakota Geological Survey. 45 p.

Baltz, E. H., J. H. Abrahams, Jr., and W. D. Purtymun. 1963. Preliminary report on the geology and hydrology of Mortandad Canyon near Los Alamos, New Mexico, with reference to disposal of liquid low-level radioactive waste. Open-file Report. U.S. Geological Survey. 35 p.

Bibby, R. 1974. Hydrogeology of the Edmonton area (northwest segment), Alberta. Report 74–10. Alberta Research. 10 p.

Clark, I., and P. Fritz. 1997. Environmental isotopes in hydrogeology. Boca Raton, FL: Lewis Publications. 328 p.

Davis, S. N., and R. J. M. De Weist. 1966. *Hydrogeology*. New York: John Wiley and Sons. 463 p.

Fetter, C. W. 1994. *Applied Hydrogeology*. Upper Saddle River, NJ: Prentice-Hall, Inc. 691 p.

Freeze, R. A. and J. A. Cherry. 1979. *Groundwater.* Englewood Cliffs, NJ: Prentice-Hall, Inc. 604 p.

Phillips, F. M. 1994. Environmental tracers for water movement in desert soils of the American Southwest. *Soil Science Society of America Journal* 58:15–24.

Pupacko, A., R. J. La Camera, M. M. Reik, and D. B. Wood. 1987. Water resources data, Nevada, water year 1986. Water-Data Report NV-86-1. U.S. Geological Survey. 263 p.

Shuttleworth, W. J. 1993. Evaporation. In *Handbook of hydrology*, D. R. Maidment, ed. New York: McGraw-Hill. p. 4.1–4.53.

Stone, W. J. 1992. Paleohydrologic implications of some deep soilwater chloride profiles, Murray Basin, South Australia. *Journal of Hydrology* 132:201–23.

Stone, W. J. 1995. Preliminary results of modeling the shallow aquifier, Mortandand Canyon, Los Alamos National Laboratory, New Mexico. New Mexico Environment Department, Report NMED/DOE/AIP-95-1. 32 p.

Stone, W. J. and D. R. Brown. 1975. Rainfall-runoff relationships for a small semiarid watershed, western flank San Andres Mountains, New Mexico. *New Mexico Geological Society Guidebook,* 26th Field Conference, p. 205–12.

Stone, W. J. and W. K. Summers. 1987. Hydrogeology and river management, Rio Grande Valley, New Mexico. In Proceedings, 31st New Mexico Water Conference, Report 219, p. 145–79. New Mexico Water Resources Research Institute.

Stone, W. J., T. Leeds, R. C. Tunney, G. S. Cusack, and S. A. Skidmore. 1991. Hydrology of the Carlin trend, northeastern Nevada—a preliminary report. Hydrology Department, Newmont Gold Company. 123 p.

Van Hylckama, T. E. A. 1980. Weather and evapotranspiration studies in a saltcedar thicket, Arizona. Professional paper 491-F. U.S. Geological Survey, 51 p.

CHAPTER 7

Hydrologic Impact of the Geologic Setting

In most cases, the geology of a region exerts considerable control on its hydrologic phenomena. As used here, *geologic control* means the role of the geologic setting in influencing hydrologic features, conditions, or processes. More specifically, stratigraphy, geologic structure, and geomorphology impact the occurrence, movement, and quality of surface, soil, and ground water. An awareness of such controls and how to recognize them is invaluable in formulating the conceptual hydrogeologic models needed to solve today's water-supply and pollution problems. A single chapter can hardly do justice to this fascinating topic. However, the general overview that follows should provide a starting place for interpreting the relationship between the geology and hydrology of a study area.

STRATIGRAPHIC CONTROLS

A major element of the geologic setting is its stratigraphy. The sequence, lithology, and geometry of strata in an area all may control local or regional hydrologic phenomena.

Sequence

The variability of materials in the geologic column is an important source of control. This applies to surface water and ground water alike. For example, where a stream cuts down through a horizontal sequence of strata, the location of gaining and losing reaches will be controlled by the permeability of the different materials in the sequence. Also, if there is little variety, say more or less uniformly permeable material exists throughout the geologic section, unconfined ground-water conditions will prevail. If permeable material lies above less permeable material, conditions are favorable for perching ground water. The reverse, that is, less permeable material above permeable material, or alternating layers of permeable and impermeable material, leads to confined or artesian ground-water conditions, as in the now classic example of the Dakota Sandstone (Figure 7-1).

Lithology

The type of material or rock present in the saturated zone plays an important role in its hydrologic behavior. First, it determines whether the material is an aquifer or

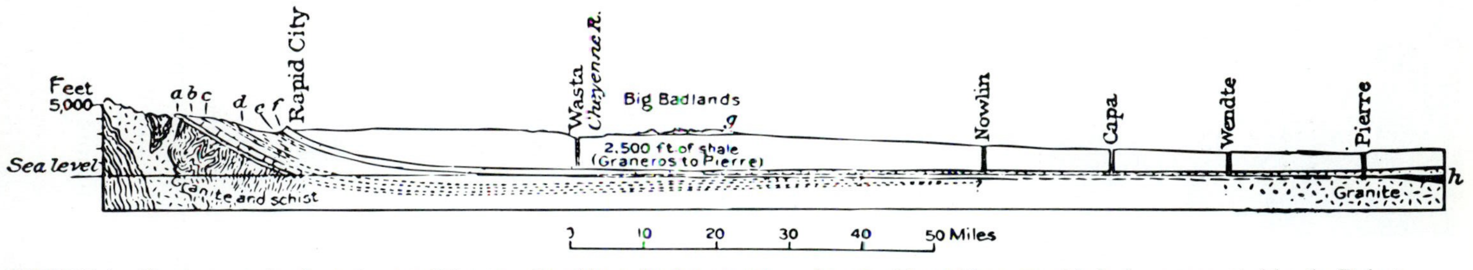

FIGURE 7-1 Classic example of artesian conditions produced by a dipping sequence of permeable and impermeable beds, as presented for the Dakota Sandstone by Darton (1918, Figure 2).

FIGURE 7-2
Intergranular porosity and permeability of clastic sedimentary materials vary with grain size and sorting. This photo shows basin fill material, Rio Grande valley, central New Mexico.

aquitard. Next, it determines the type of porosity and permeability available. If the material is granular (Figure 7-2), porosity and permeability are likely to be primary, whereas, if it is crystalline, porosity and permeability are likely to be secondary (Figure 7-3). Further, if the granular material is well sorted (for example, sugar sand or pea gravel), porosity and permeability will be greater than if it is poorly sorted (for example, silty sand or sandy gravel). When there are two well-sorted granular materials, the one with the larger grains will have the greatest permeability. Since pores in the coarser medium are larger, more water lies beyond the influence of friction at the pore walls and flow is easier.

Geometry

The shape and extent of a geologic deposit is the result of both depositional and post-depositional processes. The original dimensions may be modified by erosion before the overlying material is deposited, as will be discussed later. The geometry of a deposit strongly influences its hydrologic significance and behavior. More specifically, aquifer geometry can impact both ground-water occurrence and movement.

In the case of occurrence, only wells drilled within the boundaries of productive saturated material will be successful. This fact is especially significant where there are sharp contrasts in geologic materials. Narrow, linear sandstone bodies (such as channel or offshore-bar deposits) may be the focus of important ground-water occurrences, especially if they are surrounded by finer-grained material, such as shale. The same may be said of esker deposits surrounded by till in glacial terrain.

Additionally, ground-water movement may be influenced by the presence and orientation of such sand bodies. Since movement is easier within the more permeable material, ground-water flow direction in such aquifers is generally parallel to their length. For example, flow would be along offshore-bar, channel, or esker deposits, where they are the aquifers.

FIGURE 7-3 Fracture porosity and permeability of crystalline rocks may be enhanced by dissolution (when below the water table). This photo shows gypsum deposit, central New Mexico. Openings are up to 8 feet tall.

ROLE OF UNCONFORMITIES

As noted by Meinzer (1923), unconformities can also have hydrologic significance (Figure 7-4). Such breaks in the geologic record influence hydrologic conditions in four ways:

1. by juxtaposing materials of different hydraulic properties,
2. by determining the continuity or extent of hydrostratigraphic units,
3. by directing water movement, and
4. by controlling the potential for ground-water recharge or discharge.

The specific role the unconformity plays depends on its type, as well as the kind of geologic material above and below it.

Disconformities

A disconformity (an unconformity along which there is erosional relief) may provide all four types of control.

First, like all unconformities, a disconformity may juxtapose materials of different lithology and thus different hydraulic properties. If the material above the contact is

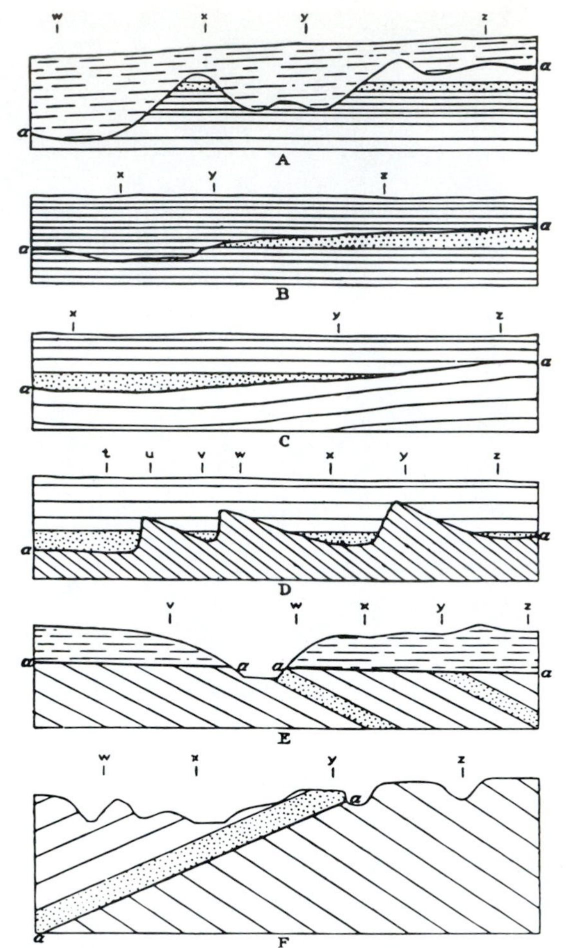

FIGURE 7-4 Hypothetical cross sections showing hydrologic impact of unconformities. Consider the water-resource potential or success of wells located at the positions indicated by the various letters. (From Meinzer 1923, Figure 55.)

permeable, but that below it is not, the result will be perched water bodies. For example, water is often perched in alluvium that overlies bedrock. This condition, coupled with the erosional relief on disconformities, often results in the development of localized, highly productive water bodies. For example, buried valleys along the disconformity between unconsolidated alluvium or glacial outwash and the underlying bedrock contain significant ground water. Such features are a prime target for ground-water exploration, especially in the glaciated portions of the United States and Canada. Where the materials above and below the contact are both permeable, the result is simply a thickened aquifer. Examples of this are thick fluvial sequences with erosion surfaces between depositional packages and sections where dune-sand deposits are overlain by permeable fluvial sandstone or vice versa. By contrast, if the material below is permeable, but that above is not, confined conditions may result. For example, the fine-grained deposits of the Chinle Formation (Triassic) provide confinement of ground water in the underlying karst San Andres Limestone aquifer (Permian) along the southern margin of the San Juan Basin in northwest New Mexico (Stone and others 1983).

Second, the erosion responsible for the relief that characterizes disconformities may also have significantly altered the hydrostratigraphic sequence. More specifically, this erosion may involve merely interrupting the lateral continuity of the unit just below the contact, reducing its extent, or removing it entirely. The consequences differ, depending on whether the material eroded is an aquifer or aquitard. If the material is an aquifer, the erosion determines the distribution of productive material and, ultimately, ground-water movement. Consider the impact of erosional truncation of marine sandstones prior to further deposition. For example, the availability of ground water in the Gallup Sandstone (Cretaceous) over a large portion of the San Juan Basin is controlled by this unit's erosional pinch-out into the Mancos Shale. In other words, ground water is only readily available where the aquifer has not been removed (west of the pinch-out). If the material beneath the disconformity is an aquitard, the erosion may have diminished its ability to confine the water below it.

Thirdly, the relief created by the erosion at disconformities may also provide controls on ground-water movement. For example, flow may be restricted to the localized aquifers laid down in the paleotopographic lows along the disconformity (Figure 7-4). Also, flow direction may be controlled by the orientation of an erosional pinch-out on a disconformity. For example, ground-water in the Gallup Sandstone, being unable to move into the Mancos Shale, turns at the erosional pinch-out and flows parallel to it toward discharge zones (Figure 7-5).

Finally, the erosion and subsequent deposition responsible for disconformities also may have enhanced the potential for recharge or discharge of ground water. Ero-

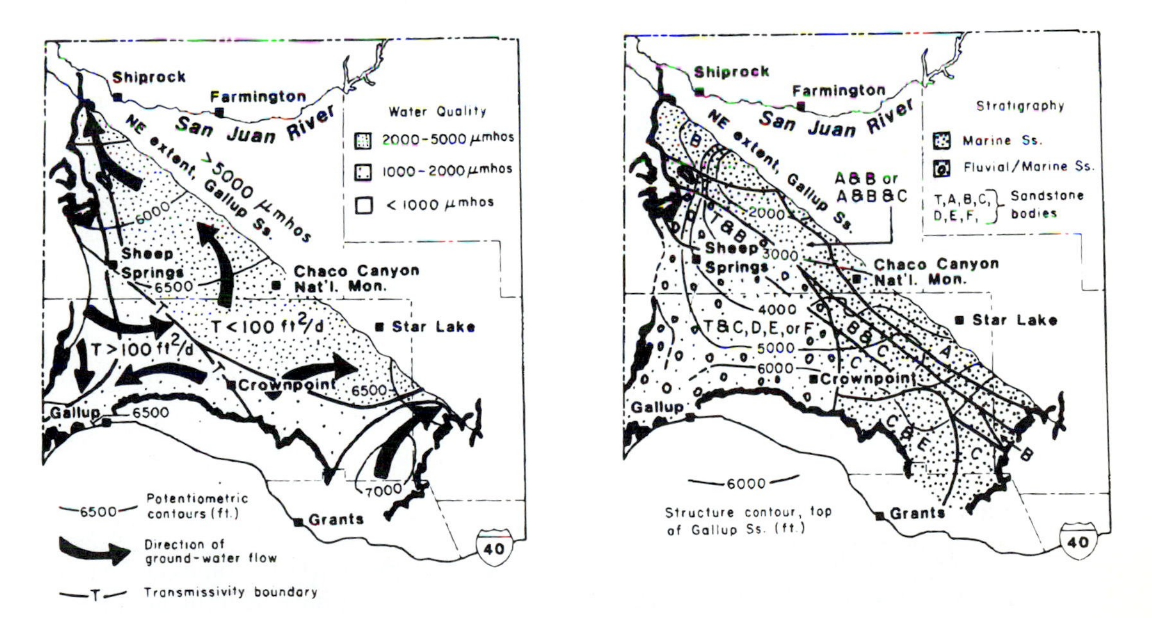

FIGURE 7-5 Maps showing the relationship between hydrology and geology for the Gallup Sandstone, San Juan Basin, northwestern New Mexico. (See Figure 2-6 for location.) Ground-water flow is parallel to the erosional pinch-out of this aquifer. (From Stone 1981, Figure 6.)

sion may have caused permeable material to subcrop at the contact. If the material overlying the contact at such places is also permeable, water may freely move down or up across it. By contrast, if erosion caused impermeable material to lie at the contact or such material was deposited above it, the opportunities for recharge and discharge are diminished.

Angular Unconformities

An angular unconformity (one in which strata below have an angular relationship to the contact, but strata above are parallel to it) provides similar controls to those associated with disconformities, since there is obviously erosion at such contacts, as well. The difference is the presence of tilted strata. In elevated locations, such as domes or uplifts around basin margins, the upturned attitude of beds beneath angular unconformities may enhance the potential for their recharge. In settings at some distance from recharge areas, the tilting of beds beneath the unconformity may enhance discharge of ground water to overlying strata.

Nonconformities

Nonconformities (unconformities between igneous or metamorphic rocks and sedimentary rocks or deposits) also control hydrologic phenomena. Although the regional water table may lie within the crystalline rock (if fractured) beneath these unconformities, such material is generally less permeable than the sedimentary materials overlying them, and water can be at least temporarily perched along such surfaces. Because there is also erosion along such surfaces, localized water bodies may develop, as they do for disconformities. Similarly, the direction of ground-water movement may be controlled by the orientation of the erosional relief. Due to the low permeability of the crystalline rock beneath them, recharge is usually minimal along nonconformities.

CONTROL BY GEOLOGIC STRUCTURES

In addition to the strata and unconformities present, the way in which they have been modified by structural processes is also an important aspect of the geologic framework of an area. Folds and faults, as well as fractures and joints, also contribute to the control of hydrologic phenomena.

Folds

Anticlines, synclines, and monoclines (and their larger counterparts, domes and basins) can exert considerable control on the hydrologic system. They have an influence on both surface- and ground-water phenomena.

In the case of surface-water systems, the presence of folds influences what drainage pattern develops, thus dictating the topographic fabric of an area (Figure 2-2). By presenting a variety of rock units to the stream bed, eroded folds may influence the distribution of gaining and losing reaches of a channel (Figure 7-6).

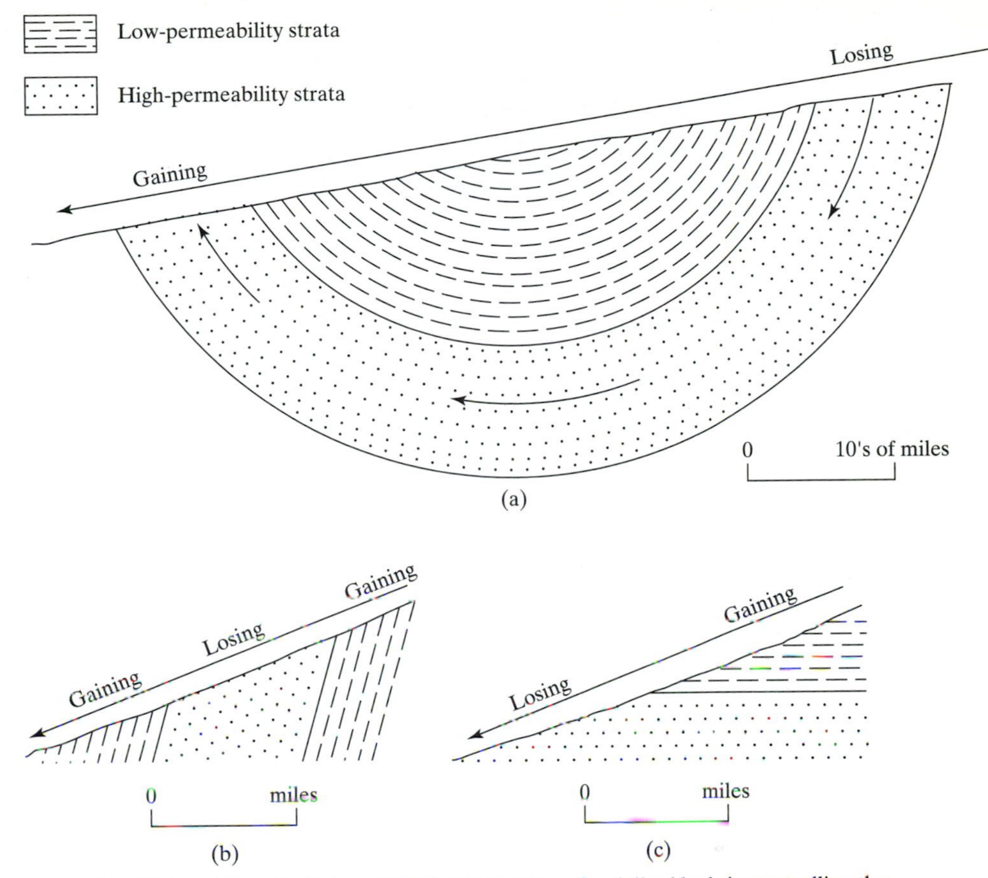

FIGURE 7-6 Hypothetical cross section indicating role of tilted beds in controlling the location of losing reaches of a stream.

Folds may also control ground-water phenomena. Fracturing associated with the crests of folds may be an important source of porosity and permeability. However, this varies with the type of fold. Fractures associated with the crests of anticlines are the result of tension and are usually open; thus they may be the focus of enhanced porosity and permeability. By contrast, fractures associated with the crests of synclines are due to compression and are often closed, resulting in low porosity and permeability. In some cases, the dip of strata in folds or basins determines ground-water flow direction. In other cases, topography may exert a greater influence (Figure 7-7). Dip, together with the appropriate stratigraphy, is responsible for classical artesian conditions (see Figure 7-1).

Faults

Like folding, faulting may also impact both surface- and ground-water phenomena. The impact depends on the nature of both the material and the fault.

In the case of surface water, faults may control the location of gaining and losing reaches as well as the flowing segments of intermittent streams. Consider the situation

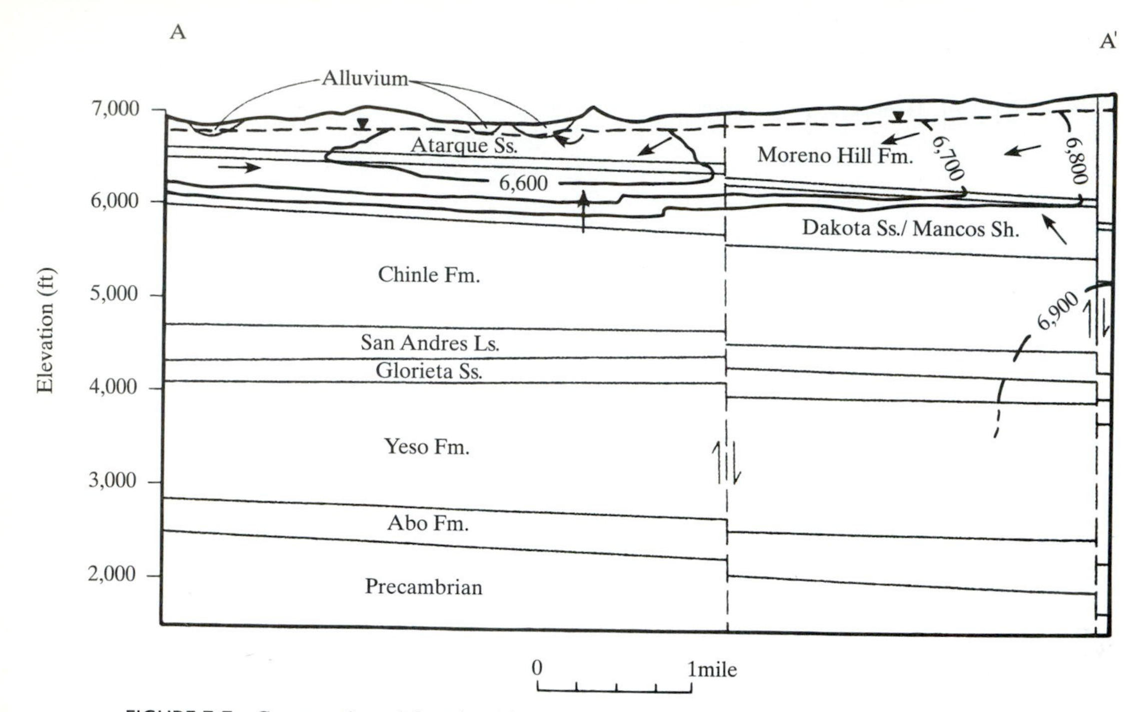

FIGURE 7-7 Cross section of situation where ground-water flow is controlled by topography rather than structure (dip), Nations Draw area, west-central New Mexico. (From McGurk and Stone 1987, Figure 3.)

where a fault is perpendicular to a stream channel and has juxtaposed permeable material against impermeable material. If the permeable material lies on the upstream side of the fault, transmission loss is abruptly reduced at the fault. If the stream bed is dry but there is underflow, water cannot readily move through the impermeable material beyond the fault, so is forced to the surface and the stream flows beyond that point (Figure 7-8a). Alternatively, if the permeable material lies downstream from the fault, flow may be reduced abruptly or soak in completely at that point (Figure 7-8b).

A fault juxtaposing permeable and impermeable material and oriented perpendicular to ground-water flow direction exerts control on water movement. If the permeable material lies on the upgradient side of the structure, the fault acts as a barrier to ground-water flow. If the permeable material is very thick and lies on the downgradient side of the fault (as at the boundary between basins and ranges), ground water may cascade down the fault (Figure 7-9). In such a case, the fault marks the position of an abrupt change in water-table depth and a sharp increase in hydraulic gradient.

Where materials on the opposite sides of a fault are the same, or have similar hydraulic properties, the fault may have no hydrologic impact (Figure 7-10). However, if the materials are brittle, the fault may locally enhance porosity and permeability. In such cases, the fault may act as a conduit, providing not only the pathway, but also the direction for ground-water movement.

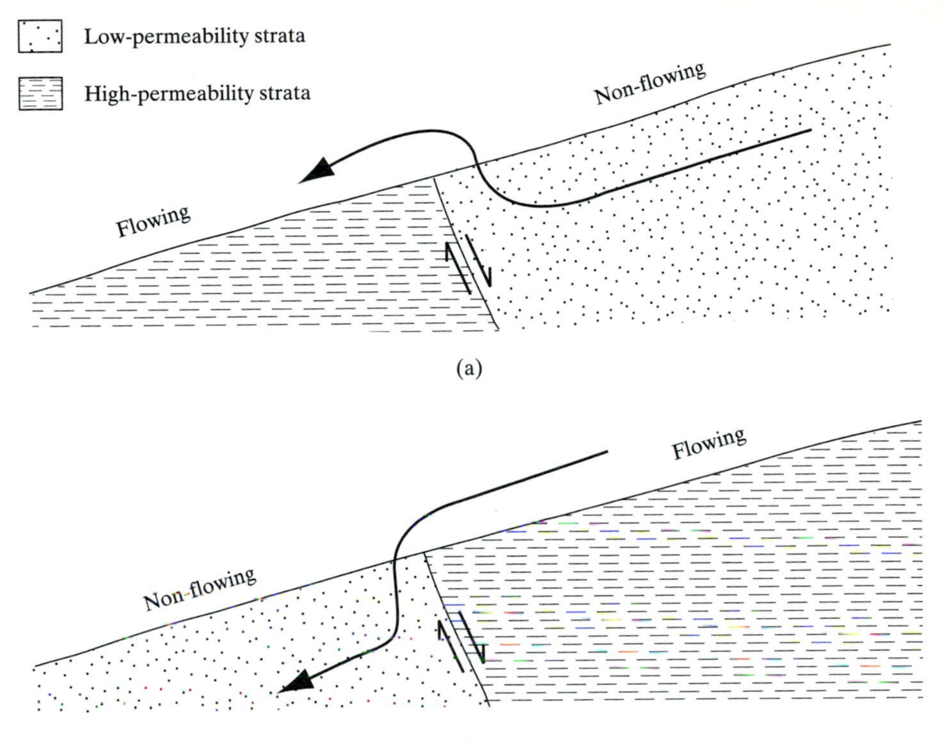

FIGURE 7-8 Hypothetical cross sections showing fault control of intermittent stream flow: (a) conditions where flowing reach is downstream of fault, (b) conditions where flowing reach is upstream of fault.

FIGURE 7-9 Schematic cross section of typical basin-and-range conditions where water cascades down a fault.

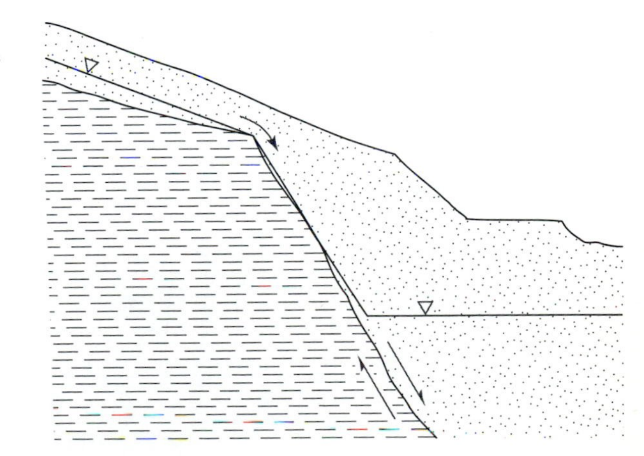

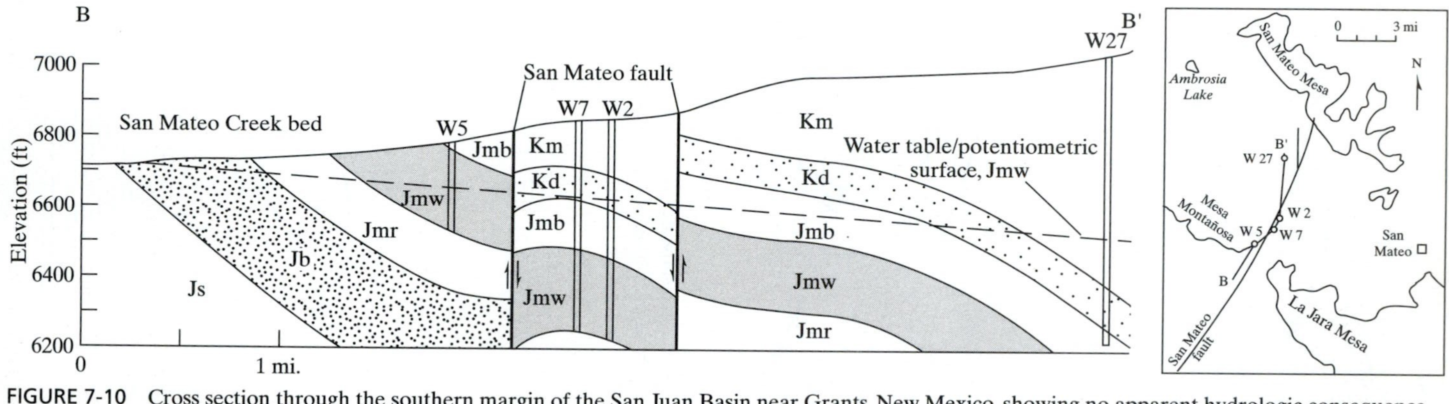

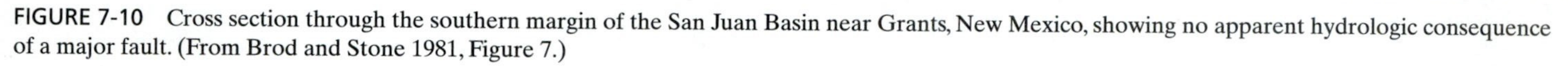
FIGURE 7-10 Cross section through the southern margin of the San Juan Basin near Grants, New Mexico, showing no apparent hydrologic consequence of a major fault. (From Brod and Stone 1981, Figure 7.)

Fractures and Joints

More or less vertically oriented breaks in geologic materials along which there has been no movement have a hydrologic impact as well. Like folds and faults, fractures and joints influence both surface-water and ground-water characteristics.

Fractures and joints are essentially the result of stress relief. When overlying or adjacent rocks are removed by erosion, stress on the rock changes. In regions of essentially flat-lying strata, such stress relief causes horizontal fractures to develop beneath valley floors in response to arching and causes vertical fractures to form along valley or canyon walls in response to tensile forces.

In the case of surface water, joints may be the focus of losing or gaining reaches of streams, depending on ground-water conditions. Joints may also exert a major control on the course of a stream. For example, the course of the River Murray in South Australia can be shown to be strongly influenced by jointing (Figure 7-11).

As with faults, the hydrologic influence of joints on ground water depends on the nature of the material in which they occur. In nonbrittle material, such as unconsolidated sediments or poorly indurated sedimentary rocks, the joints will not be open, and they may have little hydrologic impact. This is especially true in the case of fine-grained materials. In brittle rock, however, joints are more likely to be open and contribute to its porosity. If the joints are interconnected, they may also provide permeability. As with faults, the orientation of the joints exerts control on ground-water flow direction. Joints may also serve as preferred pathways for infiltration, and ultimately recharge. The most transmissive part of an aquifer in the Appalachian Plateaus of West Virginia was found to be the result of stress-relief fractures (Wyrick and Borchers 1981).

The hydrologic impact of fractures and joints may not be long-lasting because they may eventually be filled in by various materials. Clay may be washed in from the surface or form in place due to chemical weathering in the presence of water that seeps down along the joint. Lacking this, joints may become plugged by minerals precipitating out of the water (for example, calcite).

INFLUENCE OF GEOMORPHOLOGIC FEATURES

Hydrologic phenomena can also be controlled by what is generally categorized as geomorphology. These sources of control include topography, soils, karst, and related features, as well as petrogenic features.

Topography

The elevation, regional slope, and topographic relief in an area impact most parts of the hydrologic cycle. These physiographic parameters influence surface-water, soil-water, and ground-water phenomena alike.

Morisawa (1968, p. 20) noted,

> "Altitude and orientation of the basin, shape and ground slope of the watershed, relief, rock type and soil mantle, and geologic structure are all important elements in determining the hydrologic characteristics of a river system."

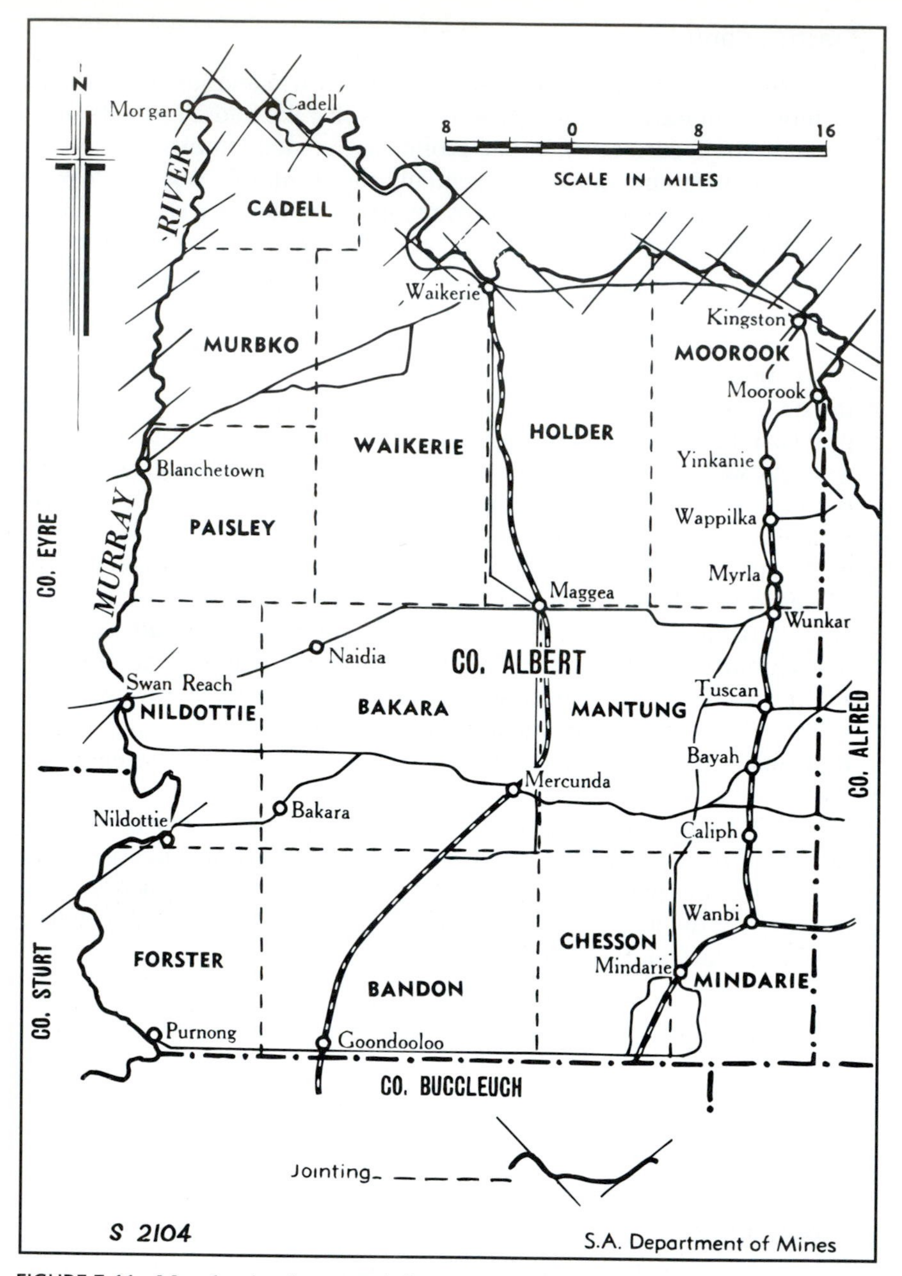

FIGURE 7-11 Map showing the strong influence of jointing on the course of the River Murray, South Australia. (From O'Driscoll 1960.)

Because streams are often largely responsible for the topography of an area, it is not quite correct to think of topography as controlling surface-water phenomena. Rather, there is a symbiotic relationship between streams and topography. For example, precipitation, thus runoff, varies with elevation. In fact, this relationship has been used to estimate runoff in ungaged areas (Moore 1968). Integration of drainage and the development of a drainage pattern makes optimum use of the prevailing slopes provided by various geologic processes (folding, faulting, volcanism, glaciation, mass wasting, etc.).

Topography also influences soil-water phenomena. The main impact is on soil-water content, which generally varies with landscape setting. On slopes, for example, precipitation, snowmelt, etc., run off rapidly, thus having little opportunity to soak into the soil. By contrast, more water is available to the soil in flat areas. This occurs for two reasons. First, there is little runoff from such areas. Second, if the flat area lies adjacent to slopes, additional water may be contributed by runon.

Ground water is also influenced by topography. It is often stated that the water table is merely a subdued version of the ground surface. Since precipitation is greater at higher elevations, uplands or mountains are generally recharge areas. Topography controls ground-water movement in that subsurface water flows from higher (recharge) areas to lower (discharge) areas. Some spring types, such as those associated with depressions, abrupt changes in slope, contacts between permeable and impermeable materials, and the erosional truncation of artesian aquifers, are the result of topographic conditions as well (see Figure 6-9). The general degree of dissection of a region (incision by streams) also controls ground-water depth. Water in highly permeable material may be as deep as the deepest canyon or valley in the area. In any case, ground water is generally deep beneath mesa tops or hills and shallow beneath adjacent valley floors.

Soils

The type of soil in an area also provides some control of the hydrologic system. While soils only indirectly impact surface-water phenomena, they can directly affect soil-water and ground-water conditions.

Between storms or runoff events, stream flow is maintained by release from bank storage. The rate and duration of this release depends on the texture of the bank material.

Soil-water content, movement, and chemistry depend to a large extent on the soil type. More specifically, soil texture and structure dictate such things as soil-water content and hydraulic conductivity. If the soil is compacted, porosity and conductivity are decreased; if it is fractured, these properties are increased. Soil structure, together with root tubes and pipes, influences soil-water movement by providing preferred pathways for flow. Soil composition largely controls the dissolved-solids character of soil water. When hardpan (calcrete, ferricrete, silcrete) is present in the soil, downward percolation of soil water is hindered. Nonetheless, water gets through. Values obtained during studies in Australia and the American Southern High Plains range from 0.06 to 1.25 mm/yr (Stone 1985).

Since ground water is replenished by soil moisture from the vadose zone, the amount, rate of movement, and chemistry of soil water has a bearing on ground-water phenomena as well. The rate and quality of ground-water recharge varies with landscape setting, a main parameter of which is soil type. For example, the long-term average rate of recharge under soil associated with dune sand in the Murray Basin of South

Australia was found to be low (0.06 mm/yr) and the water very saline (chloride content approximately 15,000 mg/L), due to the extraction of water, but not salt, by native vegetation (Allison and others 1985). By contrast, recharge rate beneath an alluvial-fan soil in New Mexico was found to be fairly high (1.02 mm/yr) and the water quite fresh (chloride content of 89 mg/L), due the permeable nature of the material (Stone 1992). An additional factor in these cases was vegetation (evapotranspiration), but that is controlled by soil type as well.

Karst and Related Features

Collapse features and depressions of various types may also exert control on the hydrologic system:

- dissolution (karst) features in limestone and evaporite terrain;
- hydrocompaction features in deposits at the mouths of mountain canyons, at landfills, in spoil at mines or in fill at construction sites;
- collapse of abandoned underground mines; and
- depressions formed by wind.

All can have an impact on surface-water, soil-water, and ground-water phenomena alike.

Surface water may be captured by such features. For example, karst features can be responsible for the disappearance of runoff, interrupted streams, and underground streams (in caves). Determining the connection between surface and ground water in karst areas has often required tracer studies. Swallow holes, due to the compaction of low-density materials upon the addition of water (Figure 7-12) or to mine-roof collapse, similarly disrupt surface-water flow. Eolian depressions (blowouts) are local internal-drainage features and collect surface runoff.

Such features may also impact soil-water conditions. Water content may be increased by infiltration of the captured surface water. This increase in water input supports a similar increase in soil-water movement. Because the water captured by such features is largely unaltered precipitation, soil water remains fairly fresh.

FIGURE 7-12 Hydrocompaction collapse hole in spoil at the Navajo Coal Mine, northwest New Mexico. Scale card is 6.5 inches long.

The impact of karst features on soil- and ground-water phenomena is greatest if the features are relatively young and open. A ground-water recharge study of a Tertiary marine limestone terrain in South Australia showed that the rate of soil-water movement differed beneath younger and older sinkholes (Allison and others 1985). More specifically, recharge beneath a younger (secondary) sinkhole was found to be 60 mm/yr, but the value obtained for an older (primary) sinkhole was the same as for the undisturbed setting (lacking any sinkholes): 0.1 mm/yr. The lower rate of recharge in the older sinkhole is due to the lack of openwork: it was plugged by deposition and precipitation of calcite.

The countless blowouts on the American Southern High Plains temporarily hold water after rainstorms, thus have been referred to as "playas." Like sinkholes, these features have also been shown to be the focus of enhanced ground-water recharge (Stone 1990).

Materials subject to karst processes have some of the highest ground-water production rates known. This is due to the excellent open-work created by the dissolution of limestone and gypsum. Both the porosity and permeability of such materials are enhanced by karst processes.

Petrogenic Features

Masses of crystalline rock resulting from igneous activity or certain bodies of indurated material from sedimentary processes also exert control over hydrologic phenomena. Igneous features include the products of both intrusive and extrusive processes. Igneous intrusions include batholiths, stocks, dikes, or sills, etc. Igneous extrusions include volcanoes, lava flows, ash-fall layers, etc. Sedimentary intrusions would include such things as sandstone dikes and salt domes. Other features of a sedimentary origin would include breccia pipes and organic reefs.

The presence of such features in an area can influence surface-water runoff, as well as the recharge, flow, discharge, and quality of ground water. Craigg and Stone (1983) found that the location of springs in an area on the eastern side of the San Juan Basin in New Mexico was controlled by the presence of volcanic plugs and dikes. Ground water is perched both on and in basalt flows in the fill of the Espanola Basin near Los Alamos, New Mexico (Purtymun 1995).

An awareness of geologic controls on local or regional hydrology is essential for making sound conceptual models and wise water-resource or environmental decisions. To supplement the overview presented here, locate and digest a case history on a setting similar to that of your study area. Of particular value would be a study where geologic controls of the hydrologic system were discussed.

REFERENCES

Allison, G. B., W. J. Stone, and M. W. Hughes. 1985. Recharge in karst and dune elements of a semiarid landscape as indicated by natural isotopes and chloride. *Journal of Hydrology* 76:1–25.

Brod, R. C., and W. J. Stone. 1981. Hydrogeology of Ambrosia Lake-San Mateo area, McKinley and Cibola Counties, New Mexico. Hydrogeologic Sheet 2. New Mexico Bureau of Mines and Mineral Resources.

Craigg, S. D., and W. J. Stone. 1983. Hydrogeology of Arroyo Chico-Torreon Wash area, McKinley and Sandoval Counties, New Mexico. Hydrogeologic Sheet 4. New Mexico Bureau of Mines and Mineral Resources.

Darton, N. H. 1918. Artesian waters in the vicinity of the Black Hills, South Dakota. Water-supply paper 428. U.S. Geological Survey. 64 p.

Hillel, D. 1971. *Soil and water—physical principles and processes.* New York: Academic Press. 288 p.

McGurk, B. E., and W. J. Stone. 1986. Conceptual hydrogeologic model of the Nations Draw area, Catron and Cibola Counties, New Mexico. Report to Salt River Project, Phoenix. New Mexico Bureau of Mines and Mineral Resources. 104 p.

Meinzer, O. E. 1923. The occurrence of ground water in the United States—with a discussion of principles. Water-supply paper 489. U.S. Geological Survey. 321 p.

Moore, D. O. 1968. Estimating mean runoff in ungaged areas. Water resources bulletin 36. Nevada Department of Conservation and Natural Resources. 39 p.

Morisawa, M. 1968. *Streams—their dynamics and morphology*. New York: McGraw-Hill. 175 p.

O'Driscoll, E. P. D. 1960. The hydrology of the Murray basin in South Australia. Bulletin 35, 2 vols. Geological Survey of South Australia.

Purtymun, W. D. 1995. Geologic and hydrologic records of observation wells, test holes, test wells, supply wells, springs, and surface water stations in the Los Alamos area. Report LA-12883-MS. Los Alamos National Laboratory, New Mexico. 339 p.

Stone, W. J. 1981. Hydrogeology of the Gallup Sandstone, San Juan Basin, northwest New Mexico. *Ground Water* 19(1):4–11.

Stone, W. J. 1985. Recharge through calcrete. Memoires. Vol XVII. pt. I, p. 395–404. International Association of Hydrogeologists.

Stone, W. J. 1988. Recharge at the Cal-West Metals site, Lemitar, Socorro County, New Mexico. Open-file report 340. New Mexico Bureau of Mines and Mineral Resources. 17 p.

Stone, W. J. 1990. Natural recharge of the Ogallala aquifer through playas and other non-stream settings, eastern New Mexico. p. 180–192. Texas Bureau of Economic Geology, Geology/Regional Hydrology, Blackwater Draw and Ogallala Formations.

Stone, W. J. 1992. Estimating contamination potential at waste-disposal sites using a natural tracer. *Environmental Geology and Water Science* 19(3):139–145.

Stone. W. J., F. P. Lyford, P. F. Frenzel, N. H. Mizell, and E. T. Padgett. 1983. Hydrogeology and water resources of San Juan Basin, New Mexico. Hydrologic report 6. New Mexico Bureau of Mines and Mineral Resources. 70 p.

Wyrick, G. G., and J. W. Borchers. 1981. Hydrologic effects of stress-relief fracturing in an Appalachian Valley. Water-supply paper 2177. U.S. Geological Survey. 51 p.

PART III

SYNTHESIS

Information on the geologic setting and hydrologic system of a study area must be synthesized into a conceptual hydrogeologic model. To be useful to others, the findings of hydrogeologic studies must be presented in a sound technical report.

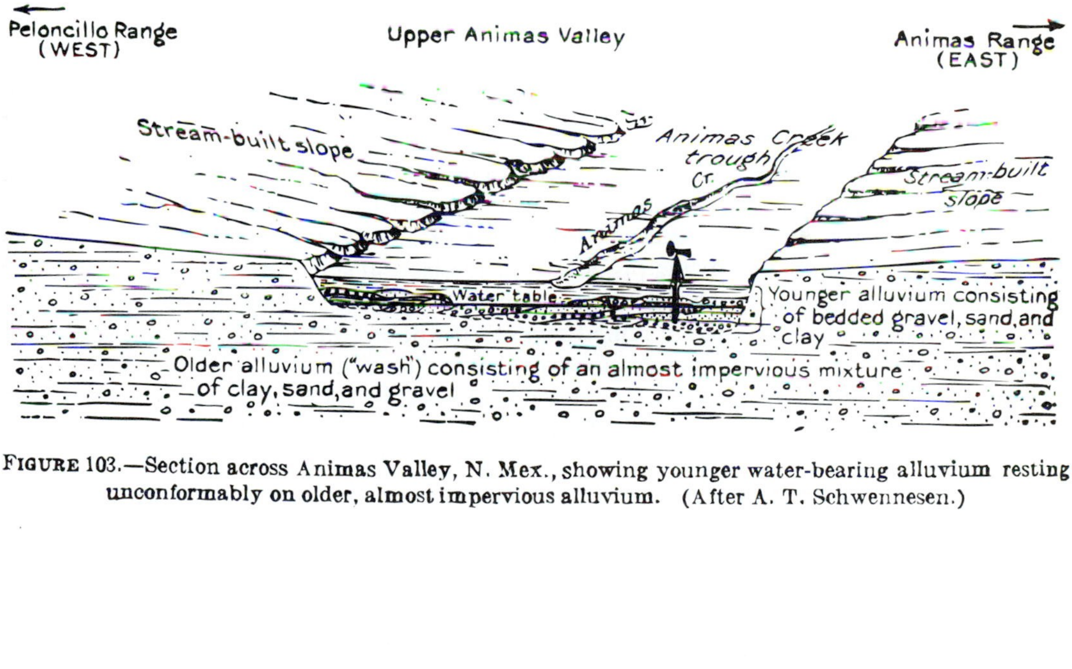

FIGURE 103.—Section across Animas Valley, N. Mex., showing younger water-bearing alluvium resting unconformably on older, almost impervious alluvium. (After A. T. Schwennesen.)

CHAPTER 8

Conceptual Hydrogeologic Models

The conceptual model is the end product of all the hydrogeologist's efforts. In the previous chapters we have looked at how to characterize the geologic setting and the hydrologic system, how various geologic features control hydrologic phenomena, and finally how the various geologic materials measure up as aquifers. All that remains is to put all this together. With this background, it is now possible to formulate a sound conceptual model of the hydrogeologic system.

IMPORTANCE

A sound conceptual hydrogeologic model is essential regardless of the type of hydrologic study being conducted. Such a model is the starting place in water-resource, waste-disposal, and remediation studies alike. It is not only premature, but wasteful to locate wells in such studies without first synthesizing what is known about the setting. For example, water-supply wells drilled without understanding area hydrogeology may be placed where the aquifer is thin or missing altogether, the aquifer is present but not very productive, or the aquifer contains water of poor quality. A sound conceptual model is also critical in characterizing proposed waste-disposal sites. Only some perception of the setting can ensure that all the geologic materials and structures present are evaluated. Finally, wells located at remediation sites without the benefit of a sound conceptual model may not adequately monitor what is intended.

Sometimes, owing to the size or complexity of hydrologic systems, numerical models are used to simulate current or future conditions (see chapter 14). As has often been pointed out, such computer models are only as good as the conceptual hydrogeologic models upon which they are based. Thus it is essential that hydrologic modelers also be able to realistically conceptualize the relationship of the hydrologic system to the geologic framework in an area, even with limited information, time, and resources.

SCALE

Conceptual models vary with the scale of the study or size of the area selected. At the regional scale you might deal with hydrogeologic provinces. At a subregional or areal scale you are normally concerned with aquifers. At the local scale different hydrogeologic domains may be described within aquifers. Alternatively, you may conceptualize the study area in terms of flow systems.

More than one of these scales may be addressed in a given study. Stone and others (1991) considered all these scales in conceptualizing the hydrogeology of the Carlin Trend, a gold-mining district of northeastern Nevada. The following abbreviated descriptions, excerpted from their report, serve as examples of the role of scale in conceptual models.

Regional

Two distinct hydrogeologic provinces are recognized in the Great Basin (Thomas and others 1986; Mifflin 1988): Low-permeability noncarbonate rocks dominate the western part of the region, and a thick, highly permeable carbonate-rock section underlies the eastern part. In the low-permeability rock province, volcanic, granitic, and clastic sedimentary rocks prevail. These contain or yield little water and form barriers to ground-water movement. Flow is further controlled by geologic structure and circulation is often restricted by basin-boundary faults. In the carbonate-rock province, thick sequences of Paleozoic limestone and dolostone with high secondary permeability dominate. Ground-water circulation between basins is common and can be extensive (Kirk and Campana 1988). Flow direction is not necessarily controlled by topographic or structural boundaries. Water may flow from the fill of one basin, through the carbonate rocks in the mountain block, and into the fill of the adjacent basin (Mifflin 1988).

Subregional/Areal

At the subregional or areal scale, hydrostratigraphic units are the major elements of a conceptual model. In the Carlin Trend, five aquifers are recognized, based on permeability, degree of confinement, and ground-water flow direction. These occur in a stacked sequence with no regional aquitards between water-bearing units. Thus, water in one aquifer is in contact with that of the adjacent aquifer. These include, in descending order as they would be encountered in drilling, alluvium (Quaternary), basin-fill sediments (Tertiary), volcanic rocks in the fill (Tertiary), siltstone (Paleozoic), and carbonate rocks (Paleozoic). It might be possible to simplify this to three aquifers, lumping the two Tertiary materials and the two Paleozoic rocks, pending more detailed information.

Local Domains

Hydrologic properties may also vary within a given unit or aquifer. This variation is most notable at the local scale where structure or mineralization and alteration have modified regional porosity and permeability. Such modifications are particularly apparent in the vicinity of the ore deposits. As many as seven domains have been recognized in one of the mine pits in the Carlin Trend. These differ mainly as to rock unit, degree of alteration, and position in the pit, as well as position relative to major faults and each other.

Flow Systems

In the Carlin Trend, three flow systems were recognized (Figure 8-1): perched, shallow unconfined, and deep semiconfined (locally geothermal). Although the deep semiconfined system is more or less synonymous with the carbonate aquifer, the other systems do not directly correspond to aquifers, but rather to position, depth, and hydrologic behavior. Aquifers are normally described separately elsewhere; here you should focus on ground-water movement.

Most studies will not require conceptualization at this many scales. Nonetheless, the site hydrogeology should be characterized at all scales appropriate for the problem at hand. Such glimpses of the hydrogeology at various scales give an immediate impression of the overall system.

COMPONENTS

The elements of an ideal conceptual hydrogeologic model are outlined in appendix C. Such a model consists of four components: geology, surface water, soil water, and ground water. Formulating a conceptual model mainly involves describing the geologic framework from a hydrologic point of view and the perceived or suspected interaction of the surface, soil, and ground water with this framework, as well as with each other. It is essential to do this for the study area, at least, and possibly for the region.

Geologic Component

The matrix within which the hydrologic system operates must be described, based on the information compiled for characterizing the geologic setting. There are several major questions to answer with text or figures.

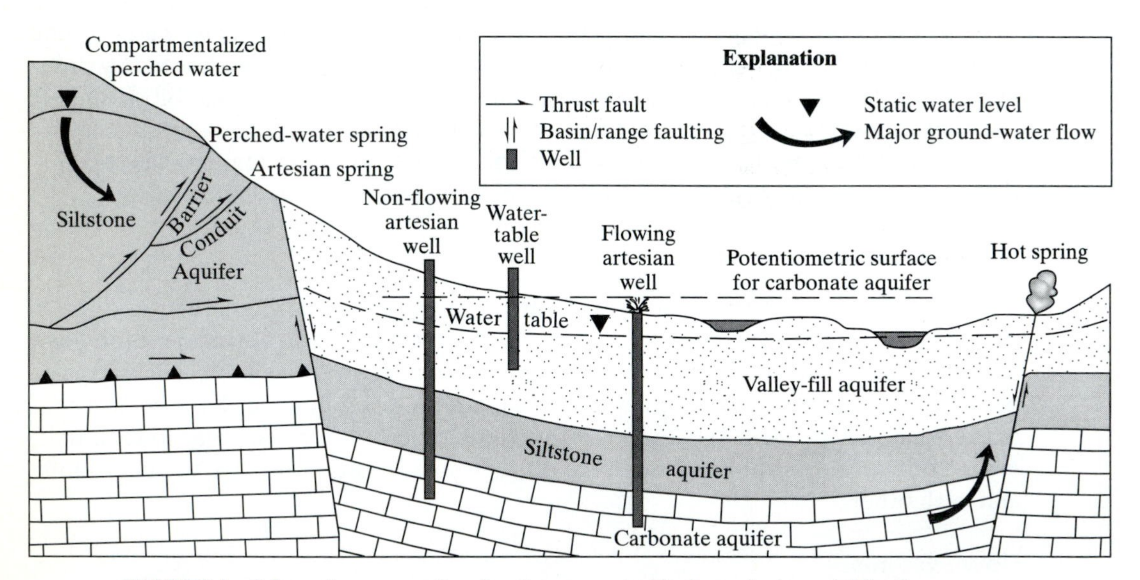

FIGURE 8-1 Schematic cross section showing conceptual hydrogeologic model for the Carlin Trend, Nevada, including major flow systems: perched, shallow unconfined, and deep semiconfined.

1. What is the stratigraphic sequence?
2. What do the units consist of?
3. What are the dip and strike of the strata?
4. What major geologic structures are present?
5. What are their extent and character?

Also, stratigraphic units should be reassigned to hydrogeologic units, based on their hydrologic behavior/function.

Surface-Water Component

The most obvious component of the area's hydrologic system is its surface water: streams and lakes. The conceptualization of the surface-water part of the hydrogeologic model focuses mainly on the water coming into or leaving the system via streams and lakes. More specifically, several questions must be answered.

1. What is the major surface-water feature in the region?
2. How does surface water in the study area relate to it?
3. Which streams, reaches of streams, or lakes are gaining or losing?
4. Does this vary seasonally?
5. What rates and volumes are involved?
6. What is the fate of the water gained or lost?

The conceptual hydrogeologic model is more concerned with the physical than the chemical aspects of the system. Nonetheless, the quality of the surface water gained or lost, as well as major changes in hydrochemistry across the area, are important and should also be described. Because some constituents, especially metals, are associated with sediments in the stream, the suspended load and its chemical make-up are an important part of the conceptual surface-water model.

Soil-Water Component

Lying between the surface water and the ground water in the study area is the soil water. Although the least well-known component of the hydrologic system, thus often the most weakly characterized in the conceptual model, it plays an important role and deserves some thought. Various questions may be answered.

1. How thick is the unsaturated zone?
2. What kind of material is present there?
3. Do preferred pathways exist?
4. What is its moisture content?
5. Is there seasonal variation?

6. What hydraulic conductivities are typical?
7. What are long-term recharge rates based on natural tracers?
8. Have there been any major land-use changes?
9. What impact did this have on vadose-zone processes?
10. Are any contaminants present in the soil water?

As for surface water, the focus is on the physical aspects of the unsaturated zone. However, the quality or chemistry of the soil water is also important, especially since this water ultimately recharges ground water.

Ground-Water Component

Depending on the number of wells in the area, various attributes of the water in saturated zones beneath the surface may be fairly well known. In any case, ground water is usually a major component of conceptual hydrogeologic models. This is important because underground water is a significant source of water worldwide. In the more arid regions, it is often the sole source of water. Of special interest are the answers to several questions.

1. Where is the recharge area?
2. What is the recharge mechanism?
3. What is the recharge rate?
4. What are the aquifers?
5. What kind of deposits are they?
6. What is their thickness and extent?
7. What is their geometry and continuity?
8. What is their homogeneity and isotropy?
9. How deep is ground water?
10. Is the water perched or regional?
11. Is it unconfined or confined?
12. What is the flow direction?
13. What is the flow rate?
14. What is the discharge mechanism?
15. Where is the discharge area?
16. What is the quality of the ground water?
17. What are the geologic controls of ground-water occurrence, movement, and quality?

Because it is used for water supplies, quality is even more significant for ground water than for the other components of the conceptual model. Of special interest is any change in water quality along its flow path. This may be evidence of ion exchange, mixing of ground waters from different aquifers, or contamination. A change in water quality at a given well through time is also significant and deserves discussion. Such changes may be due to contamination from the surface by percolation through the ground or down the well annulus.

THE WHOLE

As shown in Appendix C, the overall conceptual hydrogeologic model usually has two parts: narrative and pictorial. The narrative part not only describes the four main components outlined above, but stresses their interaction. The pictorial part includes not only the various maps, cross sections, graphs, and diagrams necessary to illustrate the setting, but also those illustrations needed to show the interaction of the components with each other and with the geologic framework. The essence of the conceptual model may often be captured by a cross section (Figures 2-9 and 8-1). Sometimes the relationship between the geology and hydrology is best shown by maps (see Figure 7-5). In other cases, a block diagram or flow chart may be required to depict the system (Figures 8-2 and 8-3).

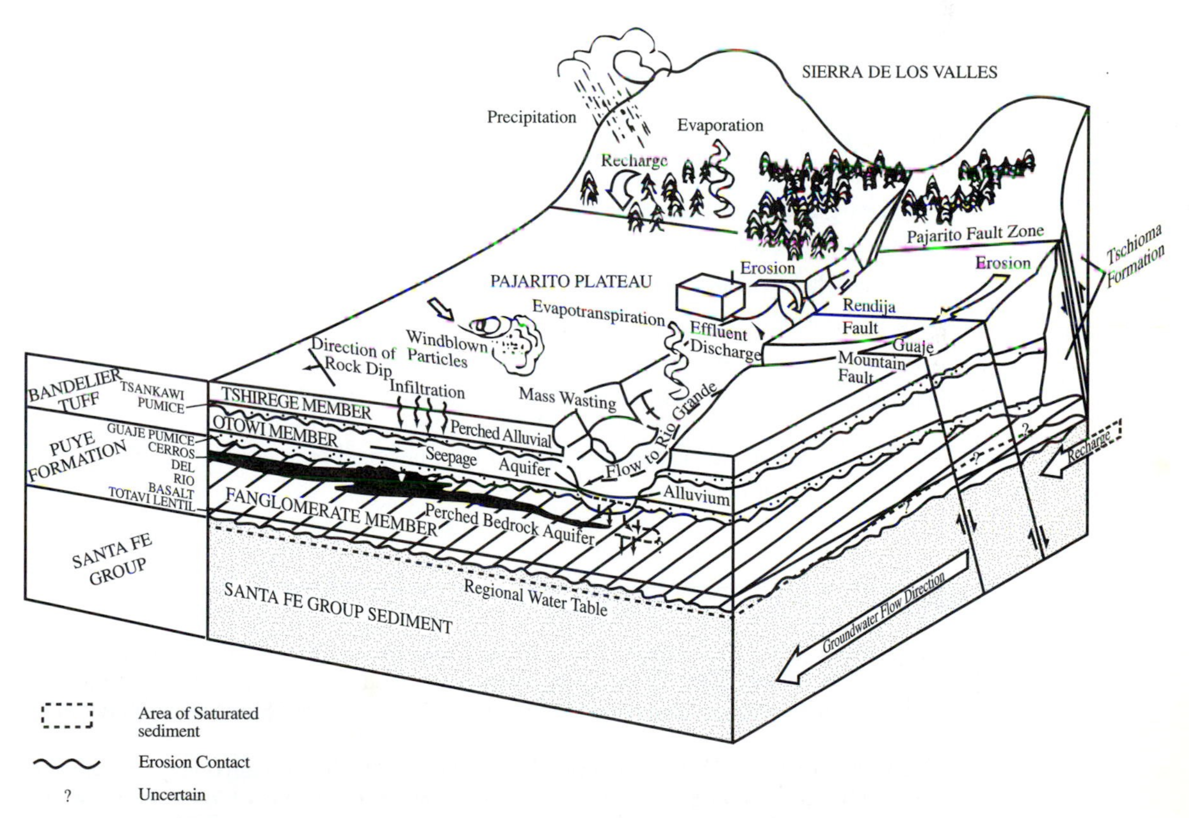

FIGURE 8-2 Schematic block diagram showing conceptual hydrogeologic model for the Los Alamos National Laboratory, north-central New Mexico. (From Aldrich 1992, Figure 4-5.)

Length

The length of the system's conceptualization depends on: (1) the complexity of the study area, (2) the amount of information available, and (3) the purpose of the study or report. If the hydrogeology of the area is complicated (e.g., involving several aquifers, a major surface-water component, and complex geology) or is the defined focus of the study, an entire report may be required to describe the system (e.g., Stone and others 1983). However, if the area is not complicated, little is known, or the conceptual model is merely intended to give a sketch of the hydrogeologic setting as background for a discussion of some other topic, much less information is necessary. The narrative part of the conceptual model may be limited to a chapter or a few paragraphs and the pictorial part limited to a few basic illustrations (e.g., Purtymun and Johansen 1974).

Fundamental Questions

Although all the the categories of information listed previously are important and should be considered, the knowledge most necessary can be gained by asking or answering 10 basic questions.

1. What is the aquifer?
2. What are its geologic and hydrologic properties?
3. How deep is the regional saturated zone?
4. Where does the water come from?
5. Which way does the water flow?
6. What is the hydraulic gradient?
7. How and where does the water come back to the surface?
8. What is the quality of the water?
9. Is there any contamination?
10. What are the geologic controls on the hydrologic system?

As many of these questions as possible should be answered by even the briefest statement of the conceptual model.

EXAMPLE

The essential aspects of the conceptual model can sometimes be stated concisely. Consider the following summary of the key elements of the hydrogeologic model compiled by Los Alamos National Laboratory for the Pajarito Plateau of north-central New Mexico (Stone and others 1993):

> Ground water occurs in three situations: (a) perched in alluvium in canyons, (b) perched in basalts and sedimentary units of the Puye [Formation] and (c) beneath the regional water table in the main aquifer. The alluvium on canyon floors is recharged by surface-water runoff. The main aquifer consists of the Tesuque Formation and the lower parts of the overlying and intertonguing Tschicoma Formation (in the western part of the Pajarito Plateau) and the

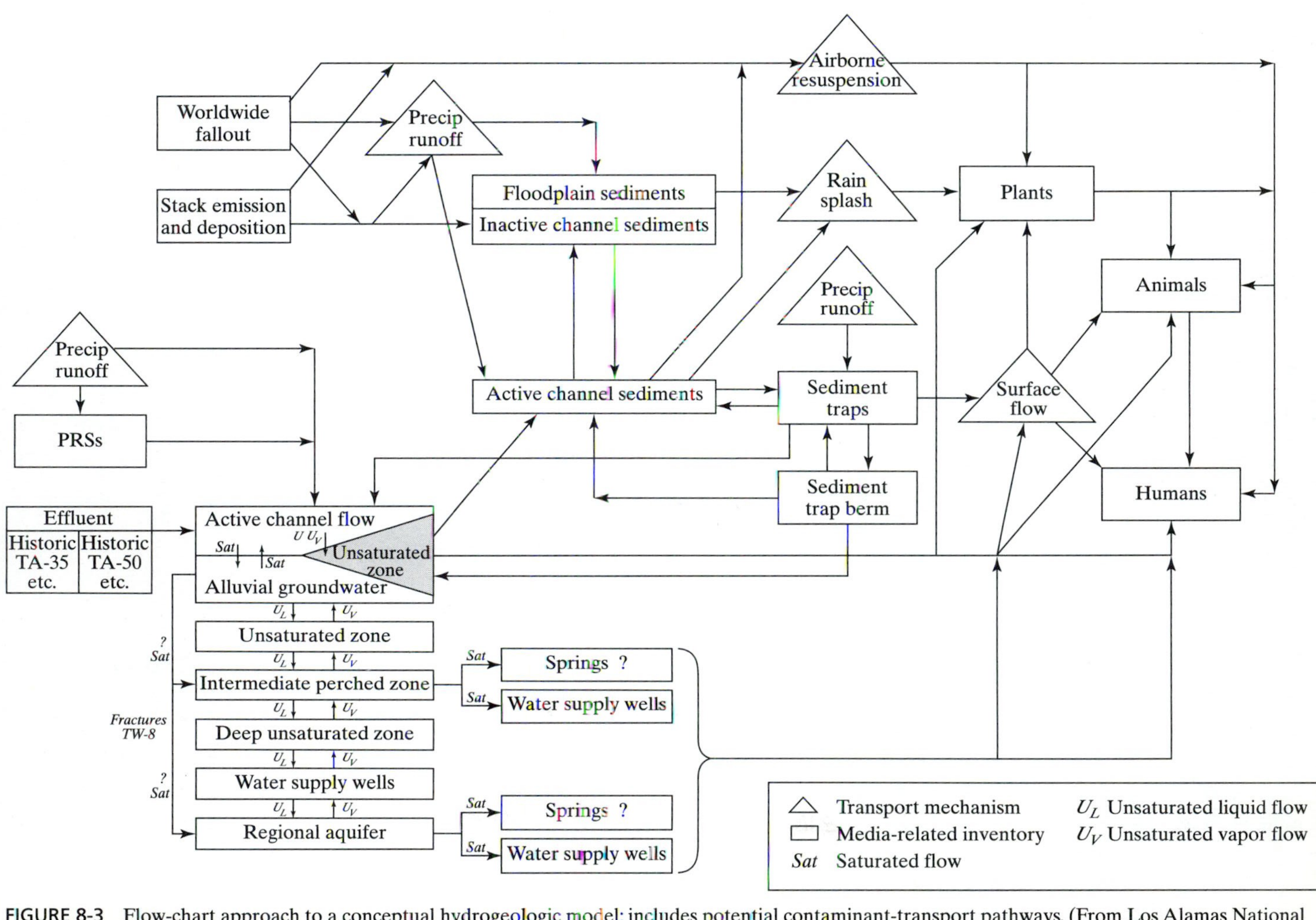

FIGURE 8-3 Flow-chart approach to a conceptual hydrogeologic model; includes potential contaminant-transport pathways. (From Los Alamas National Laboratory 1997, Figure 4.1.3-1.)

> Tesuque Formation and overlying Puye [Formation] (in the central and eastern parts of the Pajarito Plateau). Water in the main aquifer is separated from that in the shallow alluvium and other perched systems by 250–620 ft of unsaturated [Bandelier] tuff and sediments. There is little or no recharge of the main aquifer from the mesas, the shallow alluvium or other perched ground water. The main aquifer is recharged in the Valles Caldera, west of the lab. Ground water in the main aquifer flows easterly until discharging to the Rio Grande.

Although the system is complex due to multiple aquifers and multi-unit aquifers, most of the basic questions are answered. The important aspects of both ground-water occurrence and movement are addressed.

As often happens, water quality is not included in this statement of the conceptual physical model. To be complete, however, the conceptual model should address the relationship between the geologic setting, the hydrologic system, and its hydrochemistry. This relationship is generally discussed in a separate section. Several basic questions must be answered here as well.

1. What are the concentrations for the various chemical constituents?
2. Is there contamination?
3. What is the background concentration for natural constituents?
4. What hydrochemical facies are recognized?
5. How do they relate to geological conditions?
6. What, if any, hydrochemical changes occur along flow paths?
7. What is responsible for these changes with distance?
8. What, if any, hydrochemical changes have occurred over the period of record?
9. What is responsible for these changes with time?

As when conceptualizing the physical system, the hydrochemical model may be presented both narratively and pictorially. Various types of illustrations are useful in the pictorial portion of the conceptual model. Piper plots (Figure 8-4) may be used to show the relationship between the major-ion content and the materials with which the water is in contact. Figure 8-5 portrays average Stiff diagrams for the Carlin Trend, Nevada.

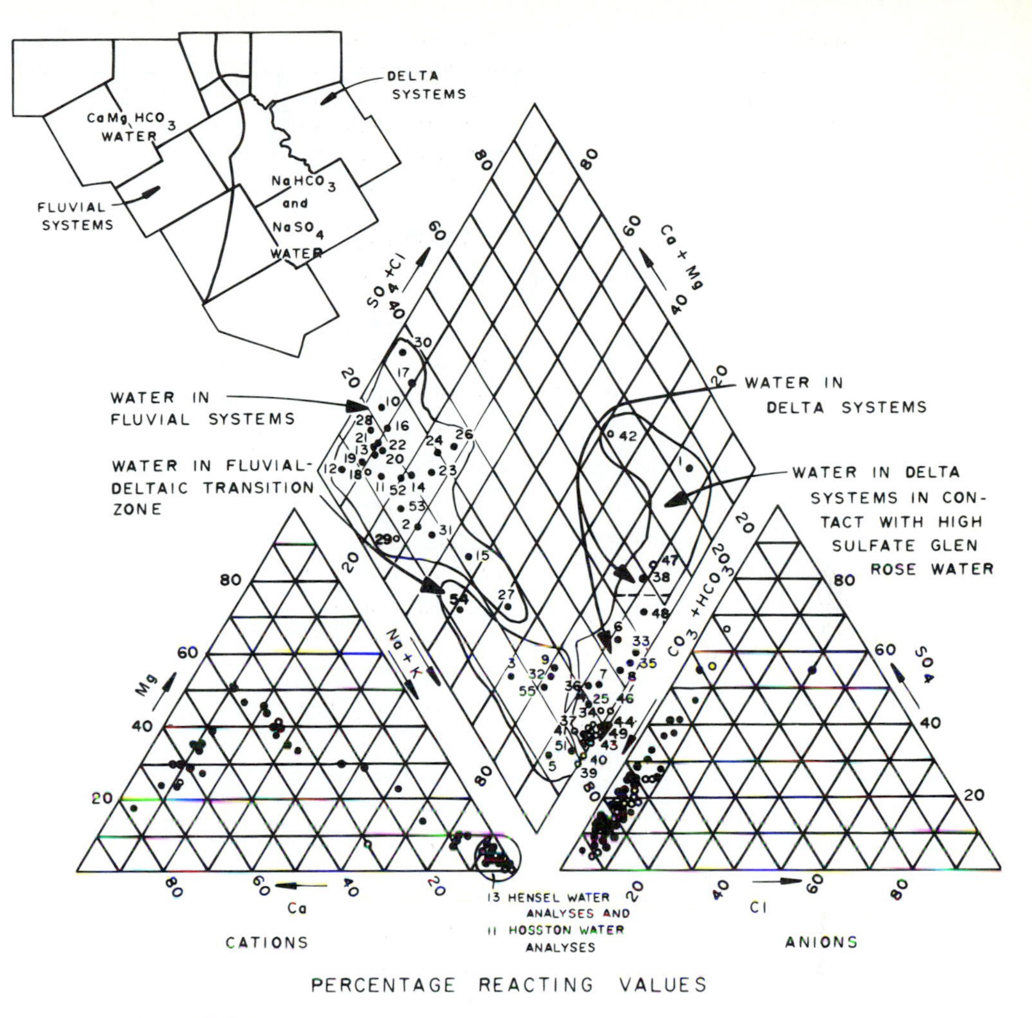

EXPLANATION

• 2 HENSEL WATER ANALYSIS AND MAP INDEX NUMBER

∘ 5 HOSSTON WATER ANALYSIS AND MAP INDEX NUMBER

FIGURE 8-4 Piper plot relating the composition of ground water and the depositional origin of two aquifers in north-central Texas. (From Hall 1976, Figure 18.)

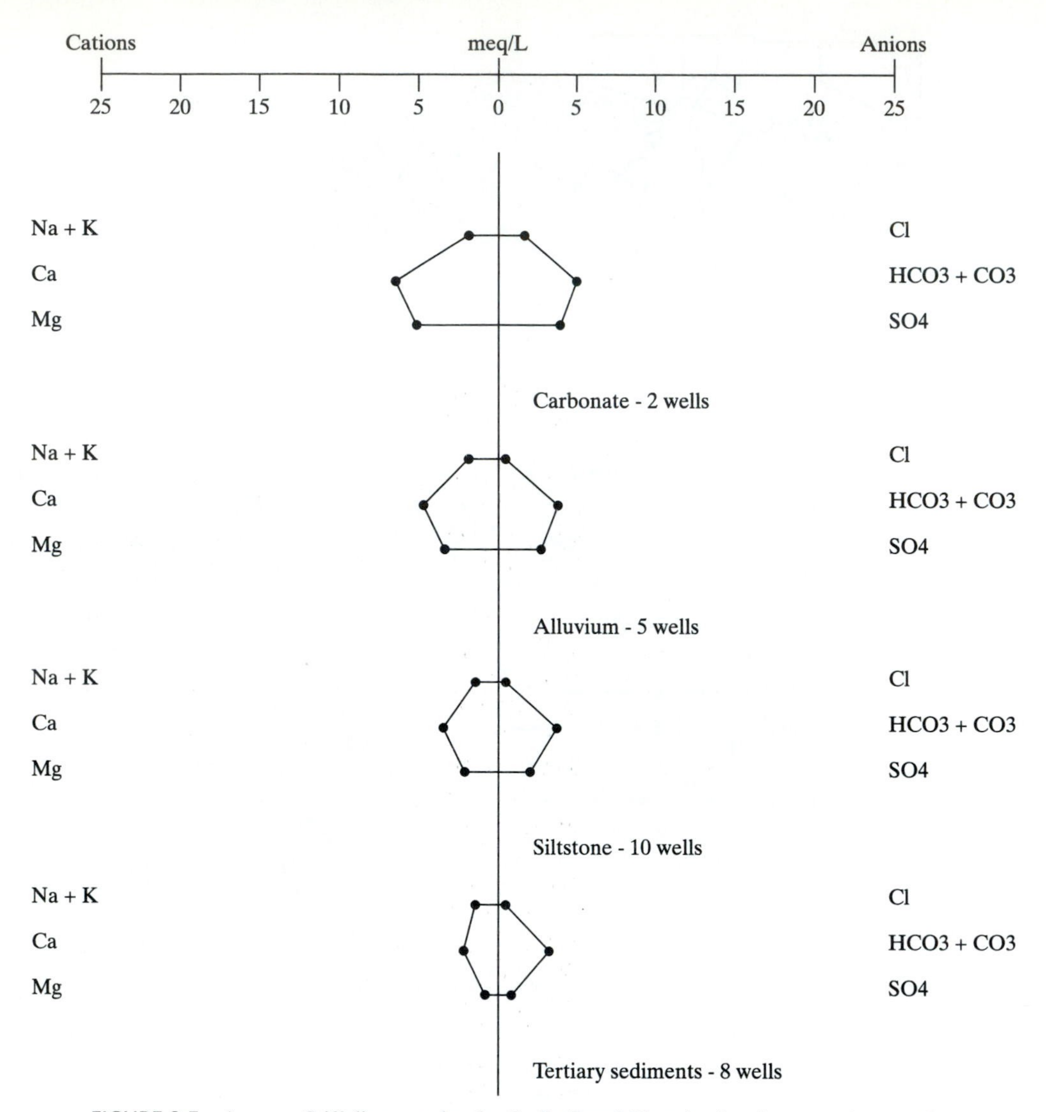

FIGURE 8-5 Average Stiff diagrams for the Carlin Trend, Nevada, showing a continuum of chemistry for waters in the various aquifers, suggesting they are in hydraulic communication. (From Stone and others 1991, Figure 29.)

REFERENCES

Aldrich, M. J., ed. 1992. RFI work plan for operable unit 1071, Environmental Restoration Program. Report no. LA-UP-92. Los Alamos National Laboratory, Los Alamos, NM.

Hall, W. D. 1976. Hydrogeologic significance of depositional systems and facies in Lower Cretaceous sandstones, north-central Texas. Geological circular 76-1. Texas Bureau of Economic Geology. 29 p.

Hoffman, L., and K. Lyncoln. (compilers.) 1992. Environmental surveillance at Los Alamos during 1990. Report LA-12271-MS. Los Alamos National Laboratory, Los Alamos, NM.

Kirk, S. T., and M. E. Campana. 1988. Simulation of groundwater flow in a regional carbonate-alluvial system with sparse data—the White River flow system, southeastern Nevada. Publication 41115. Desert Research Institute, Water Resources Center. 76 p.

Los Alamos National Laboratory. 1997. Work plan for Mortandad Canyon. Report LA-UR-97-3291. Los Alamos National Laboratory, Environmental Restoration Project, Los Alamos, NM.

Mifflin, M. D. 1988, Region 5, Great Basin. In *The Geology of North America,* Vol 0–2, p. 69–78. W. Buck, R. S. Roshein, and P. R. Seaber, eds. Geological Society of America.

Purtyman, W. D., and S. Johansen. 1974. General geohydrology of the Pajarito Plateau. Guidebook 25. New Mexico Geological Society. p. 347–349.

Stone, W. J., T. D., Davis, and D. Katzman. 1993. Initial assessment of the ground water monitoring program at Los Alamos National Laboratory, New Mexico. Report NMED/GWB-93/1. New Mexico Environment Department, Santa Fe, NM. 25 p.

Stone, W.J., T. Leeds, R. C. Tunney, G. A. Cusack, and S. A. Skidmore. 1991. Hydrology of the Carlin Trend, northeastern Nevada—a preliminary report. Hydrology Department, Newmont Gold Company. 123 p.

Stone, W. J., F. P. Lyford, P. F. Frenzel, N. H. Mizell, and E. T. Padgett. 1983. Hydrogeology and water resources of San Juan Basin, New Mexico. Hydrologic Report 6. New Mexico Bureau of Mines and Mineral Resources. 70 p.

Thomas, J. M., J. L. Mason, and J. D. Crabtree. 1986. Ground-water levels in the Great Basin region of Nevada, Utah and adjacent states. Atlas HA-694-B, 1:1,000,000. U.S. Geological Survey.

CHAPTER 9

Writing Hydrogeologic Reports

It doesn't matter how thoroughly you research an area, how clever the approach is, or how significant the findings are, if you can't clearly communicate what was done or found by means of a sound technical report. More specifically, if the organization is not logical, tables are not clear, necessary figures are not included, the project report will be of little use. Similarly, if the basic topics aren't covered, readers will not have the comprehensive picture you want them to. Of course, if there are obvious errors in even a part of the report (stylistic or factual), a shadow will be cast on the credibility of the entire work, and more than likely, your employer as well.

Consider the following excerpt from a consulting report (modified for anonymity) submitted to a state agency. Words or phrases in bold represent awkward or incorrect usage.

> Regional Hydrogeology
>
> The **outcropping geology** in the [town] area ranges in age from Triassic to Holocene. Permian to Triassic age formations **outcrop** northwest of [town] along the [river] and its tributaries. Plate 1, which shows the regional surface geology, is a copy of the [agency] geologic map of [county, state (reference)]. **Permian to Triassic age outcrops** are wide-spread across [county], however, Pleistocene to Holocene alluvium covers most of the [town] area.
>
> The older alluvium is terrace **deposits** that **consists** of **uncemented caliche**, gravel, sand, silt, and clay. It ranges from 0 to 200 feet in thickness and can yield as much as 1300 **gallon per minute** (gpm) of fair groundwater quality where the alluvium is 100 to 200 feet thick [reference]. The younger **alluvium are** probably a combination of channel deposits from the [river] and erosional material from the older terrace deposits.

There are numerous errors in both grammar and conventional usage. For example, *outcrop* is the noun; *crop out* is the verb. Also, geologic materials may crop out, but not geology. "Uncemented caliche" is a unique concept, especially to anyone who has tried to garden in the arid American Southwest. The tense and number errors should have been caught in peer review, if not when the author proof read it. In short, this is an example of how not to write.

No matter how long you have been preparing reports, you can always improve your writing skills. Everyone can. This chapter is offered as a reference for both those just starting out, who want some guidance in preparing effective hydrogeologic reports, and those with experience, but who are looking for some tips on improving their writing. This chapter does not pretend to cover everything, especially matters of style. There are numerous sources of information for style and grammar, some of which are listed in the references.

LEARNING TO WRITE

As someone once observed, "even after we graduate we're still writing theses." That's true. If you mastered the thesis (or dissertation) process, report time will go smoothly. If you didn't, or were never exposed to it, you will come to dread report time and your reports will probably reflect it.

Many hydrogeologic reports are poorly written due to inadequate training. Scientists have traditionally learned to write from their professors. This used to take place at the undergraduate level, not just in English courses, but in every assignment in other courses as well. Today, however, this duty is not taken seriously. It is apparently assumed that someone else is teaching students how to write. Admittedly, it takes a lot of time, and instructors are very busy with advising students, teaching courses, tending to committee assignments and doing research. Nonetheless, undergraduates receive little training in technical writing. For graduate students the completion of a thesis or dissertation traditionally provided the major opportunity to learn about and practice the creation of readable scientific prose. However, even that opportunity is disappearing, as a thesis is no longer required for a masters degree in some departments. Graduate students thus are in the same position as the undergraduate students. So, what can be done?

Although there is no substitute for learning the intricacies of writing in your own field at the side of a mentor, there are some measures that can be taken to improve report-writing skills. Technical writing courses are offered at most universities. If such a course was not required while you were in school, perhaps you could take one now. Your employer might even pay for it. Also, when you are reading a professional report or journal article, consider it not only a source of information on the stated topic, but as an example of technical writing as well. Published papers have been subjected to considerable peer and editorial review and most are good models of professional communication. Regardless of whether a given article is a good or bad example, you will learn something about writing. Then apply what you have learned to your own writing. Would a journal article present information the way you did? Finally, if you missed grinding through a thesis, the next best writing experience is seeing a manuscript into print. Why not polish up one of your reports for publication? A few times through that process will teach you as much about writing as completing a thesis.

GETTING STARTED

There are only three things to remember in writing a clear, comprehensive report: organization, organization, and organization. Obviously this is an overstatement, since there are many other things to consider. However, organization is the most important consideration and is required at every level: chapter, section, and paragraph. Such organization should allow you to introduce topics as they come up. Reports have topic chapters (the introduction); chapters have topic sections, sections have topic paragraphs, and paragraphs have topic sentences.

Early in the planning process, you should consider the purpose of the report. Who will be using it? What will it be used for? Are users most likely to read it cover-to-cover and base some decision on it, or are they more likely to refer to it frequently for specific data or information? Considering the answers to such questions will improve the title, organization, and contents of your reports.

Importance of Title

A report can't be written until the contents are outlined, and that can't occur until the title is selected. This may seem trivial, but the entire slant and scope of the report depends on the stated topic. Are you merely describing your specific, local study, or are you reporting on a general problem, using your study as an example? Consider the difference between these two titles: "The hydrogeology of the Tuscarora Mountains" and "The role of faults in basin and range hydrogeology." The first title sounds fairly focused and of little interest to anyone outside the area. The second has a much broader scope and promises to provide basic information of use to practically anyone. Select a title that is neither too broad nor too narrow for the material you will present. Make sure the title conveys what the report delivers. Also consider your audience. What are they familiar with? What are they most interested in? What are they likely to know already? Of the two titles given above, the first may be suitable for a company or agency report, whereas only the second is appropriate for an international journal.

Outline

Once the title is selected, turn your attention to the contents. A good way to start writing anything is to brainstorm the possible contents on some scrap paper: That is, jot down the different topics you think should be covered and those you want to cover. Don't worry about order or ranking now; just get all the topics on paper. Next, review the list to see if some topics could be combined, which aren't really appropriate (or aren't possible, even though appropriate), and what others should be added. Does the list include the kind of things you would like to see in a report of this type? When you are happy with the list, it is time to organize it. What are the major topics? Which should come first, second, etc.? What items are merely subheads for major topics? Reorganize the list accordingly. The result is not only a writing outline, but a table of contents for the report or paper.

REPORT ELEMENTS

A sound hydrogeologic report consists of various essential parts: contents, text, illustrations, tables, reference citations and lists, appendices, and possibly a glossary and an index. In the most effective reports, each of these elements has been thoughtfully designed and checked for accuracy and completeness.

Contents

There are some topics that should appear in all hydrogeologic reports.

problem
purpose
location

geologic setting
hydrologic system
conceptual hydrogeologic model
studies made
methods used
results
implications
recommendations
references

Some of these topics may be combined. For example, the first three may be covered in an "Introduction." Geologic setting, hydrologic system, and conceptual hydrogeologic model may be combined under a chapter called "Regional Setting." The topics studies made and methods used may be combined under a section called "Approach." Results, implications, and recommendations may be included in a chapter called "Conclusions." Not all of the topics in the list may be appropriate for all studies, thus can be deleted. However, the topics form an ideal list for achieving the comprehensive coverage desired.

In the section called "Problem," you might want to explain the major issue involved, as well as the portion of it you will be addressing. The purpose section explains both that of the study you are describing and that of the report itself, especially if more than one report is being released on the same general project. The report should clearly give the location relative to familiar places as well as to a land grid. The geologic setting, hydrologic system, and conceptual hydrogeologic model sections should cover the information discussed at length in previous chapters of this book. Under the topics studies made and methods, describe the general approach to the problem, what data were collected and how they were used, any special equipment involved, etc. The results should be presented in tables and figures whenever possible, but some narrative explaining what was found is inescapable. This should be done without interpretation, which is reserved for the implications ("Conclusions") section. If the data warrant further study or corrective action, some recommendations along those lines should be offered. Finally, the source of all material cited should be given in the section called "References." The brainstorming process described previously can also be used to further determine and organize what goes under each of these major headings.

In addition to these basic topics, others may be appropriate, depending on the focus of the study. For example, in the case of a water-supply report, a description of potential productivity (from pumping tests, modeling, etc.) and water quality would be obvious topics to include. If recharge is an issue, results of field and laboratory studies would be included. In remediation studies, additional space must be devoted to soils, monitoring, contaminant extent and levels, as well as plans for clean-up.

After a while you will develop a standard set of topics and format for your reports, but try not to get into a rut. Over the years I have noted that a bit of stiffness has crept into some of my colleagues' papers. Regardless of the topic, the standard headings are something like "Introduction," "Method," "Results," and "Conclusions." Although that is certainly a reasonable outline, occasionally substituting the title "Background" for "Introduction," "Approach" for "Method," and "Implications" for "Conclusions," would not mean the end of the scientific world as we know it. Be flexible, and vary the contents as dictated by the study.

Headings

It is best to keep the number of headings and subheadings to a minimum for several reasons. For one thing, it is difficult to show relative rank if there are too many subheadings. Three ranks are usually sufficient: chapter and two levels of subheads. Unless there are two or more subheads, they should not be used; in other words, there should never be just one subhead. Chapter heads are often centered, all caps; the first-order subhead is often placed at the left margin, all caps; and the second-order subhead may be placed at the left margin, first cap/lower case or placed at the beginning of a paragraph, all caps and followed by two dashes. Italics or bold type may be used to further distinguish these lowest-ranking subheads. Sometimes chapter, section, and paragraph are identified by numbers. For example, 1.0.0, 1.1.0, 1.1.1, etc., is commonly used in consulting reports. However, many find it awkward, offensive, and unnecessary.

Style

The importance of well-written, solid paragraphs cannot be overemphasized. If care is taken with them, the report will fall easily into place. Each main point under a subhead should be the topic of a paragraph. That point should be clearly stated in a topic sentence. All other sentences should support or expand on the point made in the topic sentence. If you are having trouble with a paragraph, check to see if you have clearly stated what it is about in a solid topic sentence. If you haven't, write one, and see how much easier the process goes.

Sentences themselves should be simple and to the point. Inexperienced writers have a tendency to try to say everything at once in a single gigantic sentence. This is often a series of phrases connected by *and*. It is much better to break these thoughts or points down into easily digestible statements, arranged in the most effective order.

Someone once said it is difficult to write something clearly that is scientifically incorrect. If you are having difficulty with a particular section of text, stop and consider this. Are you trying to put a round peg in a square hole? In other words, are you trying to say something that is scientifically impossible or at least improbable?

Sometimes it is difficult to say even in a lengthy section of prose what a single figure can convey almost instantly. While illustrations should not go unmentioned in the text, it is sometimes easier to only briefly summarize the information illustrations provide, and refer the reader to the figure. For example, imagine trying to compete with a hydrograph for clearly conveying the nature of water-level change at a well during a drought or streamflow during a storm. Even pages of text may not be able to present what such a figure can.

Finally, take care that the rules of good grammar and usage are followed. Although a thorough discussion of grammar is beyond the scope of this book, there are a few things to note that are not always covered by style guides.

Writers seem to particularly have trouble using non-English words correctly. More specifically, various words of Latin origin are commonly misused in hydrogeologic reports, as well as in daily language. This mainly involves mistakes regarding number: The plural form is used when the word should be singular, and singular verbs are used with plural nouns. For example, *media* is the plural for medium. Thus, the correct usage is, "this medium is . . ." and "these media are. . . ." Similarly, *data* is the plural of *datum,* so it is "this datum shows . . ." but "these data show. . . ." Additional examples include criteria, phenomena, strata, etc. Other foreign words should not be used unless they are understood. For example, it is incorrect to say "Rio Grande River" (rio is Spanish for river), "Sierra Nevada Mountains" (sierra is Spanish for mountains) or "Ojo Caliente Spring" (ojo is Spanish for spring). Various versions of Roget's *Thesaurus* (e.g., O'Conner and Pearsons 1995) give the meaning of foreign words and phrases as well as abbreviations based on them. One of these books should be consulted when you are in doubt.

As noted in chapter 3, hydrogeologists should also strive to use stratigraphic nomenclature correctly. In addition to using the correct formal name, take care to avoid confusing time and position when discussing stratigraphic conditions. Only time can be early or late; only position can be lower or upper. Thus, it is incorrect to write "Upper Cretaceous time" (upper refers to position not time) or "depth to the Early Cretaceous" (early refers to time not position). See the paper by Owen (1987) for further helpful tips on correct usage of stratigraphic terminology. Also, it is misleading to say "the Cretaceous Dakota Sandstone." Consider the construction "the red wagon." This suggests there is another wagon that is not red. The same applies to rock stratigraphic units. Such usage suggests there is another Dakota that is not Cretaceous. Therefore, it is better to put the age after the name: Dakota Sandstone (Cretaceous) or Dakota Sandstone of Cretaceous age. Finally, remember that all parts of formal stratigraphic names are capitalized. Lack of capitalization of the rank term in a name denotes informal status. Thus, a name given as Dakota sandstone is confusing. The reader does not know whether this is an incorrect reference to the well-known formal unit or a correct reference to an informal, albeit poorly named, unit.

Some other matters of style also deserve special attention. For example, beware of repetition. Check to see if the same word is used more than once in a given sentence. It should be possible to reword the sentence to avoid the repetition. Also try to conserve words. "There are" not only gets overworked, but is awkward and wasteful. Consider the sentence, "There are two wells tapping the Ogallala east of town." This may be rewritten as, "Two wells east of town tap the Ogallala." The revision is more direct and "there are" is not necessary.

It is also helpful to readers if your reports are consistent. This pertains to such things as the amount of detail, the order of presenting information, units of measure, etc. For example, if in one place you work from the general to the specific, or regional to local, you should do that throughout the report. When comparing characteristics of two or more features, sites, etc., use the same properties or units of measure and present them in the same order.

For detailed guidance as to correct usage, one may consult any of several sources. An introductory college English textbook should answer general questions. Although specific manuals do not exist for hydrogeologic writing, those prepared for geologic reports are helpful. A major source is one of the many books entitled "Suggestions to Authors," prepared by the USGS for their staff (e.g., Hansen 1991). Another geology-oriented style guide is "Geowriting," published by the American Geologic Institute (Bates 1995). Any library or bookstore should provide various other choices.

Check to see if your employer has developed a style manual or has a preference among those available elsewhere.

Illustrations

As suggested above, the old maxim, "a picture is worth a thousand words," is nowhere more true than in the case of scientific reports. The "picture" may be a line drawing or a photograph. A map, a cross section, or a graph can portray relationships far more easily than any number of well-written text pages. Although all illustrations used should be referred to in the text, they must stand on their own. That is, they must have a complete legend, as well as a concise and informative caption, so that they are understandable without reading the text. Of course, the figure should provide what the caption promises. Unfortunately, this is not always the case (Figure 9-1).

Illustrations fall into one of two categories: figures and plates. Figures are generally small enough to fit on a single page of the report or, if over sized, small enough to fold down to page size. Captions are placed at the bottom and start with the word *Figure,* followed by a number, a period, and then a brief title. The caption should employ the same terminology as used in the text reference to it. The source of the figure or any part not original to the author of the report must be included in the caption. This is usually done parenthetically at the end. The caption should also refer the reader to related figures, plates, or tables in the report that would be helpful in understanding the information or concept illustrated. The term *plate* has been applied to different things in re-

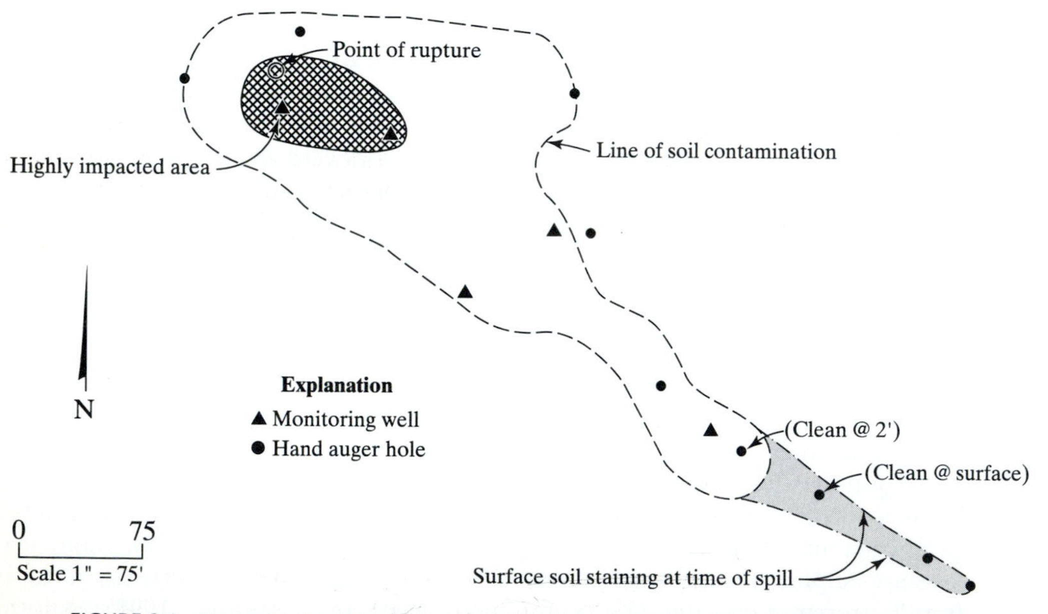

FIGURE 9-1 Example of poor text figure (modified for anonymity). Original figure upon which this is based was labeled "contour map of free-phase hydrocarbons," but it shows zones rather than contours and is not labeled correctly for either.

ports over the years. In some older reports, photographs were treated as plates, while drawings were considered to be figures. In paleontologic works, the mosaics of photos or sketches of specimens were called plates. Today, plates are generally those illustrations that are larger than page size. These are usually folded and placed in an envelope at the back of the report, or, if numerous, in a separate binding or pocket. Geologic maps, water-level maps, and any subsurface or special-theme maps fall into this category. Like figures, plates should stand on their own, having complete legends and captions. The captions appear at the bottom, following the same format as for figures. Finally consider whether all the illustrations needed are provided and whether all the illustrations provided are really needed. The reader will appreciate a good balance of the two.

It is important to be logical and follow standard conventions in the construction of illustrations. The more logical they are, the easier they are to understand. Employing counter-intuitive symbols or colors muddles the presentation. For example, red should not be used for acceptable conditions and green for unacceptable conditions. Also, avoid using a different symbol on logs, cross sections, or maps than that dictated by convention or what everyone has come to expect. For example, don't use the standard limestone pattern for sandstone on a log or cross section. Similarly, don't use a symbol for a rock type to indicate something other than rock type on a map. There are also standard colors for the periods in the geologic time table. Incorrect use of them may lead to confusion. If you are unfamiliar with what the "standard" patterns and colors are, check an introductory or field geology text. The choices available in some computer graphics software may be limited, but come as close as you can. When depth is one of two variables being graphed, assign it to the *y* axis and put zero and the *x* axis at the top for realism (as in Figure 5-4).

Tables

Tables are lists or matrices relating topics, observations, or characteristics. They, too, must stand alone. Captions are normally placed above tables, but otherwise follow the format used for figures and plates. As was the case for illustrations, include the source of the information as well as other pertinent illustrations in the report the reader should be aware of. If tabular material is too lengthy, it should be given in the appendix. If too voluminous even for that, note that supplementary data are available upon request from open files, and give their location (and possibly cost).

As was the case for illustrations, it is also important to be logical in constructing tables. For example, if showing the change in some parameter, don't use a + prefix for values to indicate a decrease and a − for an increase. In both instances, the reverse is more logical and expected. Also, if listing values separately, give the minimum value before the maximum value so that the range can be readily perceived.

References

Get into the habit of properly referencing sources of ideas that are not your own. It is not only the ethical thing to do, but also provides a reminder should you require additional information on that topic later on. The process involves two steps: briefly citing the author(s) in the text and giving a complete reference in a list at the end of the report.

The text citation usually consists of the author(s)'s last name and the year of the reference. If no author is specified in the material, use the sponsoring agency as the author (or "Anonymous") and the year (both in the text and in the reference list at the end of the report). When there are two authors, both names are given; if there are more than two authors, the first author's name is given, followed by "and others" or "et al." (Whichever form is chosen, it should be used consistently throughout the report.) Cite the original source of information wherever possible. For example, if Jones cited Smith as the source of a given water level, you should also cite Smith (not Jones). Ideally, you should consult that original reference as well. Similarly, cite the original reference, not a lexicon, as the authority for a specific stratigraphic name.

Some conceptual confusion has led to two common errors in text reference citations. Consider the usage, "Jones and Smith (1970) shows...." The first error is having the reference, presumably an inanimate object, committing the action rather than the author(s) of the reference. This is implied by the use of the singular verb. The second error is using the present tense, when whatever Jones and Smith said was in 1970 and is an act long past. Thus, all verbs in text citations should logically be in the past tense. The correct usage, therefore, would be, "Jones and Smith (1970) showed...."

No matter how computerized we get, it is still hard to beat the 3" × 5" card method of keeping track of references. Even computer bibliographies generally start out that way. In this method, a card is filled out whenever a reference is used. The format is exactly as that to be used in the report. Cards are especially useful since they can be: (1) inserted or removed as necessary during report preparation, (2) arranged alphabetically and chronologically for the final reference list, and (3) saved for use in future projects. If the card set is large, or you expect to be working in an area for an extended period, it is useful to obtain a 3 × 5 card file box (sold in any discount or office supply store) and label it for the area or project.

It is a fairly simple task to prepare the reference list for the back of the report. Simply go through the text and jot down on scrap paper the authors and dates cited (no order). Then check the list against the card set to see that all references cited are represented by a card and all references for which there is a card are in fact cited. After any discrepancies are corrected, the reference list is typed from the card set. Once the list is entered into a computer or word processor, it may be modified as needed for use in various reports on the project.

The references at the end of the report utilize a standard sequence, punctuation, and form of information. There are almost as many ways of presenting references as there are journals or employers. That is all right as long as a consistent form is used throughout the list. The form used may vary from employer to employer. Nonetheless, it should include everything necessary to locate the work in a library. It is a nuisance to try to locate a work in a periodical for which volume, number, and pages are not known.

The list at the end of the report may be called different things, depending on its content. If everything in the list was cited in the report, "References Cited" or more commonly "References" is appropriate. If the list includes more than that, "Selected References" is used. Only exhaustive lists on a subject, specifically compiled for the purpose, are considered a bibliography, thus the term does not apply to the reference lists normally generated for a hydrogeologic report.

Appendices

The appendix is the proper place for information that would be a distraction in the text. Such information includes lengthy tables, analytical results, descriptions of methods used, details on the calculation of a particular parameter, field notes, logs of cuttings, well-construction diagrams, data sheets, etc.

Each category of information is a separate appendix. If there are more than one, they are usually assigned letters: appendix A, Appendix B, etc. Each should be preceded by a cover page giving the letter designation, a brief title, and any necessary explanatory information (for example, a key to abbreviations, the units of measure, the method employed, the source of the information, etc.). If there are text illustrations summarizing the information or showing the location of samples, the reader should be referred to them by a note on the cover page.

Proofing

The last step in report preparation is checking it for typographical and factual errors. Although there is software to check spelling and grammar, that does not solve everything. There is still no substitute for a careful reading. This process should include four separate kinds of proofing:

1. checking the text (for logic, style, and usage);
2. checking figures and tables (confirming that all are cited in the text);
3. checking references (verifying that all those cited in the text, illustrations, and tables are listed at the end); and
4. checking factual accuracy of any numbers given.

Some companies/agencies have formal in-house review and proofing procedures. If yours does not, and you feel you have spent so much time with a report that you do not really see it anymore, have a colleague (or two) do this proofing.

What you have to say will vary from project to project. However, if the suggested contents topics offered here are included, the information presented is supplemented by appropriate and well-constructed illustrations, tables and appendices, as needed, and the style hints given are incorporated, the report should be effective. If you continue to apply these guidelines, every report will be better than the last one.

REFERENCES

Bates, R. L. 1995. *Geowriting—a guide to writing, editing and printing in earth science.* Alexandria, VA: American Geological Society. 138 p.

Hansen, W. R., ed. 1991. Suggestions to authors of the reports of the United States Geological Survey, 7th ed. Washington, DC: U.S. Government Printing Office. 289 p.

O'Connor, J., and E. Pearsons, eds. 1995. *Webster's school and office thesaurus.* New York: Random House. 612 p.

Owen, D. E. 1987. Commentary—usage of stratigraphic terminology in papers, illustrations and talks. *Journal of Sedimentary Petrology* 57(2): 363–72.

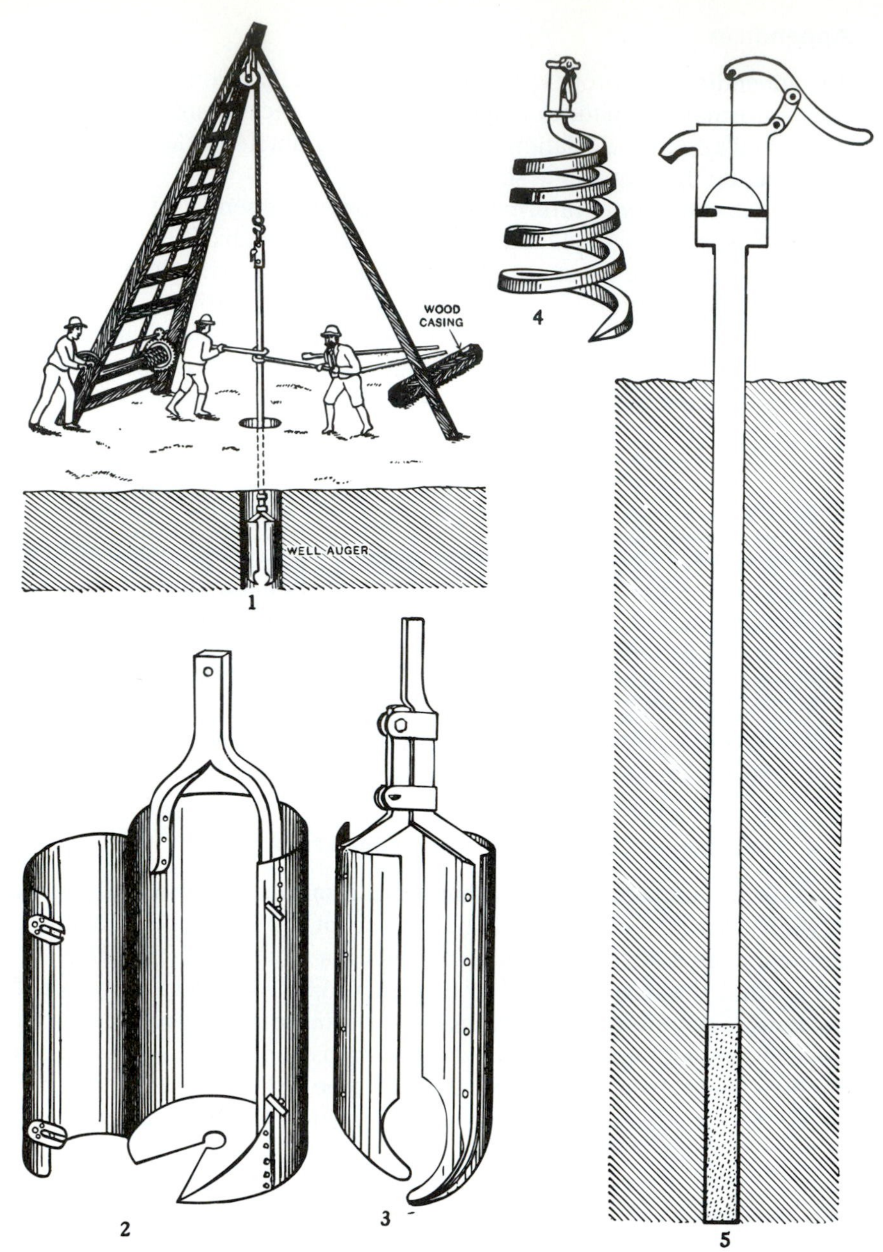

FIGURE 18.—Well-boring tools. 1, Ordinary well-boring outfit; 2, 3, well-boring augurs; 4, special auger for penetrating bowlders or rock; 5, suction pump.

PART IV

APPLICATIONS

The guidelines presented in previous chapters apply to all hydrogeologic investigations. Examples of the major types of such studies are included here to illustrate how they are applied.

CHAPTER 10

Water Supply

One of the most common applications of hydrogeology is locating or evaluating groundwater supplies. In such studies, you must define water requirements, investigate the availability of a suitable quantity of water, and determine the chemical quality of available water. Hydrogeologists may also be involved in the drilling, constructing, and testing of wells, as well as suggesting a water-treatment system.

WATER REQUIREMENTS

The suitability of a potential source of supply is often dictated by how the water is to be used. Before undertaking a water-supply investigation, it is important to consider the requirements. How much water is needed? Where is it needed? What yield or delivery rate is required? Are there any constraints on water quality?

The quantity of water required varies considerably with the type of use. As might be expected, commercial and industrial demands are the greatest. Feth (1973) reported that fabricating industries used an average of <0.5 gal per employee per day, but processing industries (including the extractive industries) used 1.2–25 gal or an average of 6.5 gal per employee per day. A New Mexico study (Sorensen and others 1973) projected that the extractive industry alone would use more than 300,000 ac/ft by the year 2000 (Table 10-1). Largest demand was projected for the fossil-fuel and copper indus-

TABLE 10-1 Projected Water Use by New Mexico Mineral Industries in the Year 2000

Industry	New Water Use (ac-ft)
Fossil fuels (processing)	117,270
Copper	52,770
Potash	42,410
Uranium	36,890
Molybdenum	33,980
Petroleum (secondary recovery)	18,000
Industrial minerals	4,330
Lead-zinc	4,060
Petroleum (drilling)	2,670
Total	312,380

Source: Modified from Sorenson and others 1973.

tries with >117,000 and >52,000 ac/ft, respectively. Projected municipal demands were the next largest. Since total human water consumption varies with population, use is calculated in terms of gallons per capita per day (gpcd). Average use for cities in the United States, was approximated 40 years ago by Leopold and Langbein (1960) as 150 gpcd. This value is still a fairly reasonable average; however, in many cities, especially in the more arid southwestern states, use has decreased through conservation measures implemented in response to dwindling ground-water supplies. Domestic demand is the lowest with strictly in-home use generally ranging from 10 to 80 gpcd. If water for landscaping is included, the value is larger.

The distance between the water source and the point of use is also an important consideration in water-supply investigations. Usually, a supply can be located at the site where it is needed; however, this is not always the case. For various reasons, a well or well field may have to be located at some distance from the site of intended use. For example, the quantity or quality of water required may only be available elsewhere. There may also be economic advantages to an off-site location, such as shallower water table or easier drilling due to the lack of a hard zone (igneous rock mass). Also, there may be no water rights associated with the use site.

Required delivery rate for the water is another concern. If the yield of the potential supply is great, the delivery system can be engineered to provide water at a wide range of rates. By contrast, if the potential source has a low yield, multiple wells may be necessary to produce the required volume and delivery rate. Regardless of how it is achieved, the supply must be reliable and continuous.

Water-quality requirements also vary with intended use, and specific constraints are associated with specific uses. Some of these constraints are imposed by regulations such as drinking-water standards. These constraints include limits for both aesthetic (e.g., total dissolved solids) and health reasons (e.g., arsenic). Other water-quality constraints are dictated by the efficiency of a manufacturing processes. There are specific tolerances for hardness and total dissolved solids (TDS) in certain industries. Whereas water with a TDS content of 200 ppm would be acceptable for the production of carbonated beverages, it would be rejected for the manufacture of rayon, acetate fiber, or paper (Swenson and Baldwin 1965, Figure 5). In addition to TDS, there may also be constraints on the content of specific constituents that may interfere with the chemistry of industrial processes.

GROUND-WATER AVAILABILITY

Determining the availability of a suitable supply of ground water requires using all the basic hydrogeologic skills and tools described in earlier chapters (chapters 2 through 8). Even if the client has conducted some preliminary studies, it is a good idea to review the available literature, consult the standard sources for basic geologic and hydrologic information, and construct a working hydrogeologic model for the study area.

The search for a suitable supply is essentially the search for a suitable aquifer. In other words, what local geologic material in the zone of saturation is capable of producing the required quantity and quality of water? If there is a county water-resource report, answering this question is a simple matter. Lacking such a local report, determine the geologic unit that other wells in the area are tapping. The best source of this information is the "aquifer" column in the county-by-county well-records database maintained by the USGS district office. If the area wells are not in such a database, determine the depth of the nearest well by consulting a driller's log filed with the state engineer or from an interview with the owner or driller. Then, using the well depth and the unit at the surface from a geologic map, together with local stratigraphy (the geologic column and unit thicknesses), the probable aquifer may be identified. If there are no nearby wells, a similar process, utilizing the regional water-level map for water depth, a geologic map and the local stratigraphy, or simply a cross section, should suggest a potential target aquifer.

Once an aquifer is identified, you must evaluate its suitability for the project. If a large quantity of water is required, low-volume or low-yield aquifers will not suffice, at least with a single well. For example, while a fine-grained material yielding a few gallons per minute is acceptable for a livestock supply run by a windmill, it will not support irrigated agriculture or a municipal supply. Also, even though an adequate quantity of water is available from a given aquifer, it may be deemed unacceptable as a supply if its quality is unsuitable. Furthermore, even an aquifer capable of providing both suitable quantity and quality of water may be rejected for economic reasons (too deep, bad location relative to point of use, etc).

WELL DRILLING

Once a target location and aquifer have been identified, a hole is made. Although very shallow water may be accessed by a simple drive point (hollow pipe with the lower portion screened), producing deeper water usually requires drilling (Figure 10-1).

There are countless drilling methods, each with its own specialized equipment. A full discussion of methods is beyond the scope of this book. However, a brief overview is provided for those new to the water-supply field. Readers interested in more details are encouraged to see one of the various specialized texts on the subject. That by Campbell and Lehr (1973) addresses drilling for oil and mineral exploration, as well as water,

FIGURE 10-1 Water-well drilling rig in operation in the Rio Grande Valley, south-central New Mexico. (Photo by John Hawley. Used with permission.)

and includes an annotated bibliography on well technology. The text by Driscoll (1986) gives an excellent summary of water-well drilling, including interesting historical notes. Another good reference is the handbook by the Roscoe Moss Company (1990).

Generally speaking, water-well drilling techniques fall into one of two broad categories: percussion and rotary. There are several specific methods within each of these categories.

Percussion Methods

Percussion drilling involves pounding or jetting a hole into the earth. Three common percussion methods include cable-tool drilling, hammer drilling and jetting. In the cable-tool method, a heavy drill string, tipped with a sharpened tool or bit, is regularly lifted and dropped into the hole by a cable. Frequently rotating the drill rods by hand ensures a round hole. Water in the hole combines with the cuttings or loosened sediments and rock fragments to form a slurry. In the unsaturated portion of the hole, water is added to generate a slurry. This mixture of water and cuttings is periodically removed from the hole with a sand pump or bailer.

In the percussion-hammer method, compressed air is used to regularly pound the bit downward, with a process much like that of a jack-hammer. In this method, compressed air is also used to bring cuttings to the surface.

Jetting is similar to cable-tool drilling, facilitated by the introduction of pressurized water on either side of a chisel-like bit. As in the conventional cable-tool method, the drill rods are rotated by hand to keep the hole round. However, shorter strokes are used to lift and drop the tools than in cable-tool drilling. Water and cuttings flow up the annulus between the drill string and the bore wall.

Rotary Methods

Rotary drilling involves penetrating geologic materials by means of a rotating bit and drill string. The bit is cooled and cuttings are removed by continuously circulating drilling fluid. There are various rotary methods, differing mainly in the type of drilling fluid used and direction of its circulation. In some methods, the drilling fluid is water or drilling mud (a specially formulated slurry), and in some it is compressed air. In conventional mud- or air-rotary methods, fluid is pumped down the inside of the drill pipe and comes up the annulus or space between the pipe and the wall of the bore. As the name indicates, the direction of flow in the reverse rotary method is the opposite of that in traditional rotary drilling. That is, drilling fluid moves down the annulus by gravity and then up through the pipe.

WELL CONSTRUCTION

Although an essential part of water-supply studies, well installation is not a hydrogeologic task and thus is beyond the scope of this book. Nonetheless, a brief overview is included to round out the coverage of water supply. The interested reader is directed to consult one of the numerous references on the topic (for example, *Public Health Service* 1963; Campbell and Lehr 1973; U. S. Environmental Protection Agency 1976; Gibson and Singer 1977; Anderson 1984; Driscoll 1986) for more details.

After a hole to the potential aquifer has been drilled, it must be completed so that it yields water as effectively as possible. Completion requires decisions about both construction materials and well design. Materials include the casing and the well screen, as well as annular fill and seals.

Casing

The casing may be thought of as the lining of the well. It consists of the pipes installed in the hole to keep it from collapsing or caving in and to seal off undesirable productive zones. Casing also provides a smooth surface for easy insertion of equipment (bailers, pumps, water-level probes, etc.). In practice there are usually at least two lengths of pipe: a surface or protective casing and the production or final casing.

The protective casing is a short, large-diameter pipe set at the surface to prevent damage to the bore by the bit and drill pipe during frequent trips in and out of the hole. It also serves to prevent contaminants from entering the well, both during drilling and after the project is completed.

The production casing is a longer, smaller diameter pipe extending from the bottom of the hole to the surface. Whereas surface or protective casing is generally black (carbon) steel, this casing may be made of stainless steel, plastic (PVC), or other material (teflon, fiberglass) necessary for local conditions, based on considerations of depth (strength) and cost.

Screen

The screen is the portion of the production casing containing openings to permit water to enter the well. The screen is of the same diameter and is positioned in it to be opposite the productive zone(s) in the aquifer. Openings may be provided by various means. In steel screens the form of this opening may include several types of louvres, mill-knife slots, or continuous slots (in wire-wrapped screens). Openings in plastic screen may be continuous slots (as in their wire-wrapped steel counterparts), sawed slots, or, rarely, drilled holes. The openings cannot be so large that they permit particulate matter from the aquifer to enter the well. Although the filter pack helps prevent such entry, the screen openings should be selected for the filterpack used (see annular material discussion next), or if none is used, the grain size of the aquifer material. Obviously this can only be determined on a case-by-case basis.

Since water cannot be produced from even the best of aquifers if the water level falls below the screen, they are usually made long enough to allow for drawdown due to pumping and gradual water-level decline. In multi-aquifer wells, screens are usually placed opposite all productive zones. Thus, the determination of screen length and position in the casing string is site-specific.

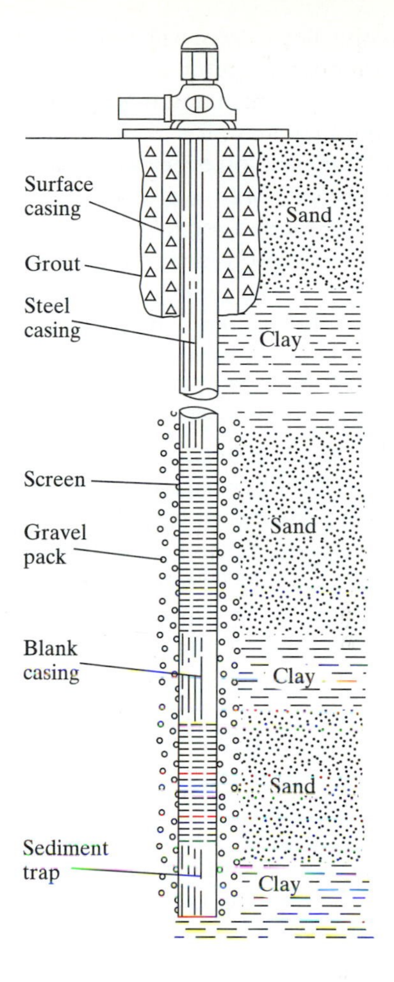

FIGURE 10-2 Generalized diagram of a water well, showing relationship of annular materials and casing. (From Heath 1983, p. 53.)

Annular Materials

Different materials are placed in the annulus, or space between the wall of the boring and the casing, for different purposes. These materials are generally intended to either promote or prevent water movement into the well (Figure 10-2).

The filter pack extends from slightly below to slightly above the well screen; because some seal material may bleed into the filter pack, the extension ensures that it does not reach the screen and enter the well. The function of this granular material is twofold: to facilitate movement of water and to prevent movement of particulate matter from the aquifer to the screen. Ideally, the material is of uniform size, shape, and composition. Grain size may range from sand to gravel, depending on the grain size of the aquifer. Textural analysis of the aquifer is required; the use of grain-size results to determine the appropriate size for the filter pack material has been fully explained by Driscoll (1986) and Aller and others (1991). A spherical shape is preferred to maximize porosity. To preserve water quality, the more inert the composition, the better. That is the reason quartz is often used.

If the screen and filter pack were sized correctly, the amount of particulate matter entering the well is minimal. However, there is a tendency to use the same size screen and filter pack on every well, without determing what is appropriate. When grain-size analysis is ignored, material may enter the well, and it is said to "make sand." At least some suspended sediment is inevitable, no matter how much effort goes into design. Thus, it is a good idea to place a blank section of casing below the screen to act as a sump, so that no sediment builds up in the screened interval.

A much different material is used to protect the quality of water produced. Because the annulus can be a conduit for transmitting contaminated water from the surface or poor-quality water from other producing zones, impermeable material is placed above the filter pack. Under artesian conditions, such material is also placed below the filter pack. Various materials are used in such annular seals, but all contain swelling clay (bentonite). The filter-pack interval is made longer than the screen; thus, if any seal material bleeds downward (or upward in the case of artesian conditions), it can not reach the screened interval and enter the well.

There may be a considerable length of annulus above the filter-pack seal. This interval need not be filled with costly material, and cuttings are sometimes used. However, as bentonite is on-site anyway, it is often used.

Contaminated water from the surface is further denied access to the annulus by installation of a surface or sanitary seal. For maximum protection, this seal consists of two parts: a grouted interval and a protective pad. The grout is generally made of concrete or neat cement (a mixture of 5 or 6 gal of water per bag of Portland cement). Sometimes bentonite is also added. The grout may merely be a continuation of the annular seal above the filter pack or may be a separate installation between fill material emplaced above that seal and the ground surface. Further isolation of the well from contamination is provided by a protective pad around the surface casing. This consists of a concrete slab around the pipe. To be effective, the pad should extend some distance out from the casing (2 or 3 ft is usually adequate) and have a mounded top to prevent ponding at the wellhead.

FINAL TASKS

Even though a bore has been drilled, water has been found, and a well has been constructed, several things remain to be done. More specifically, the well must be developed, its productivity tested, the well sterilized, a pump selected, storage provided, and water treatment considered.

Development

After the well is constructed, it should be "worked" to clean out any particulate matter that may have been incorporated during emplacement of the filter pack and to close up the contact between the aquifer, the annular material, and the screen. This may be accomplished in any way that forces water in and out of the screen, filter pack, and aquifer:

FIGURE 10-3 Pumping test under way at new supply well for the village of Magdalena, west-central New Mexico.

surging, pumping, etc. This is the same process by which a natural filter pack is produced in granular aquifers.

Testing

In order to adequately select a pump, the well yield or productivity must be known. More specifically, the drawdown or water-level drop associated with various rates of pumping must be investigated (Figure 10-3). This is determined by means of a pumping test, which involves measuring several parameters in the field (Figure 10-4):

1. the static water level before pumping is started
2. the time pumping starts
3. the pumping or discharge rate
4. the time of any change in discharge rate
5. the water levels at regular intervals during pumping
6. the time pumping stops

This book cannot provide an adequate discussion of how such measurements are made or how they are used to determine well properties. It is even difficult to cover more than the basics of pumping tests in an introductory ground-water hydrology course. Thus,

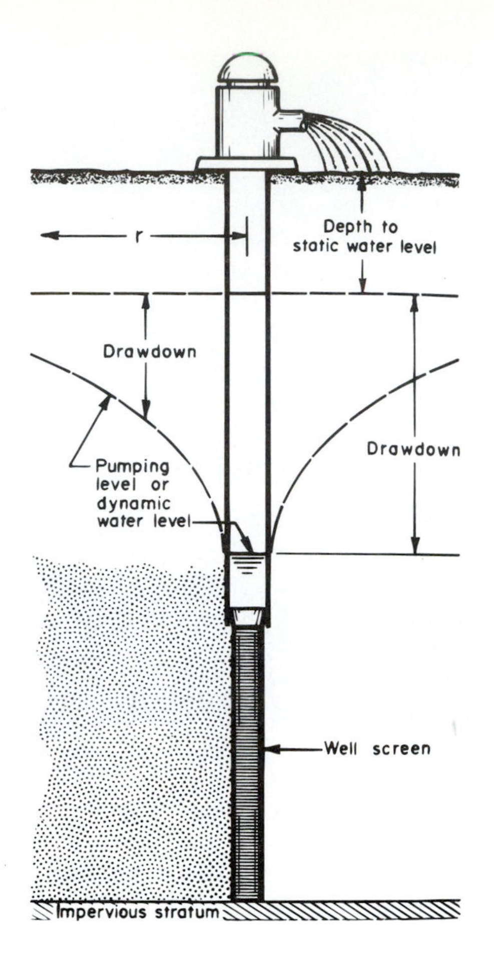

FIGURE 10-4 Schematic illustration of the parameters of interest in a pumping test. (Reproduced from Anonymous 1972 with permission from U.S. Filter/Johnson Screens.)

professional short courses devoted to the topic of aquifer analysis are commonly offered. Until you obtain such training, consult the well-hydraulics sections of standard hydrology texts and various specialized books on pumping or aquifer tests (e.g., Ferris and others 1962; Kruseman and de Ridder 1991; and Lohman 1972).

Sterilization

If the well is not periodically disinfected and construction materials are not sterilized, bacteria may be introduced during drilling and well completion. Thus, the final step in water-supply projects is well sterilization. This is commonly done with a chlorine solution. Since this agent can only treat what it comes in contact with, water in the well must be thoroughly mixed with the disinfectant. The driller should be familiar with well sterilization techniques. Driscoll (1986) gave a good summary of the necessary materials and methods.

Pump

Many specific kinds of pumps are available, and they may be classified in various ways. They may be divided into two broad groups, based on their emplacement: at the surface or within the well. While surface pumps employ suction lift, in-well pumps rely on various other means of lift.

Pumps may also be divided into two general categories, based on their yield relative to head: constant displacement and variable displacement. Both categories are suitable for pumping water from supply wells. Constant-displacement pumps produce the same amount of water regardless of head, as long as it is within their operating capacity. Examples of this category include piston (reciprocating), rotary, screw, and squeeze pumps. By contrast, the amount of water produced by variable-displacement pumps depends on the head against which they are operating. Centrifugal, jet, and air-lift pumps are in this category. The submersible pumps, commonly used, are merely underwater centrifugal pumps.

Each general category and specific type has its advantages and disadvantages. Selection will depend on site-specific conditions and requirements. Local drillers will be familiar with what works best in the area and will often include the pump in their contract. Alternatively, supply stores and manufacturers can provide information for various pump types, makes, and models.

Storage

Regardless of how it will be used, some means of storing water is needed. The type and capacity of the storage device selected depend on site-specific conditions. A common type of storage consists of a closed tank elevated to support gravity flow of the water. Alternatively, the tank may be pressurized to promote flow without being elevated. In some systems, such as those supplying mine-dewatering product to agricultural uses, a surface impoundment is used to store water.

Water Treatment

When water of unsuitable quality is all that is available, treatment is required. If the problem is minor, such as hardness, treatment at the point-of-use by simple, commercially available techniques (such as water softeners) may suffice. However, if the problem is more serious, such as contamination with organic compounds, special or large-scale treatment will be required. Driscoll (1986) provided an excellent overview of water treatment problems and techniques.

A major concern in storage (and delivery) is the protection of water quality. Thus, take care to seal any openings and avoid cross connections with systems carrying water of undesirable or dangerous quality.

CASE HISTORY

Water-supply studies may be divided into two categories, based on scale. Some assess large hydrogeologic systems, identifying and characterizing the aquifers and their water quality across an entire region (county, plateau, basin, etc.). Others describe and evaluate the ground-water availability at a specific location (mine, construction area, home site, etc.). Many excellent examples of both may be found among the publications of the USGS (Water-Supply Papers, Water Resources Investigations, etc.). Also refer to examples from your specific area among the publications of your state water agencies (especially the state geological survey and state engineer).

Rayo Hills, New Mexico

A rancher had purchased a property that was supposed to have an established livestock watering system. However, it was found that the supply was inadequate. Thus, the rancher contacted the state geological survey (New Mexico Bureau of Mines and Mineral Resources) for information on the availability of ground water in the area and the best location for a new well.

Water Requirements

The need was for a livestock supply. The well only had to be productive enough to support a windmill. Quality of the water only had to be suitable for stock watering.

Ground-Water Availability and Suitability

The ranch is located at the western edge of the Great Plains geologic province. A county ground-water report revealed that Permian gypsum deposits lie at the surface throughout the area. Although gypsum may be very productive, if there has been dissolution along fractures and bedding planes, water in contact with such rock is generally undesirable, owing to its high total-dissolved-solids and sulfate content. Nonetheless, that was all that was available, because deeper units (Permian redbeds) were not only characterized by poor-quality water, but by low yields as well. The county ground-water report showed the water quality to be within the range acceptable for stock watering and that windmills had been successfully completed in the gypsum.

Results

Thus, a well site, targeting the gypsum, was located in a valley to minimize drilling to water table and to take advantage of any recharge associated with runoff. The well was drilled by cable-tool method to a depth of 397 ft. Static water level was 300 ft. A 100-ft length of perforated pipe was used for screen. This was set in the interval 297 to 397 ft. The driller estimated productivity was at least 10 gpm and probably greater. Thus, the well was equipped with a submersible pump rather than a windmill. Water was piped to stock-watering troughs in the area.

Although this case history deals with a relatively minor single-well supply, the procedure involved is no different than would be used for a larger study. Applying the tools and guidelines presented in previous parts of the book increases the chances for success of water-supply projects.

REFERENCES

Anderson, K. E. 1984. *Water well handbook.* Belle, MO: Missouri Water Well and Pump Contractors Association. 281 p.

Aller, L., T. W. Bennett, G. Hackett, R. Petty, J. H. Lehr, H. Sedoris, D. M. Nielsen, and J. E. Denne. 1991. Handbook of suggested practices for the design and installation of ground-water monitoring wells. Report EPA/600/4-89/034. U.S. Environmental Protection Agency. 221 p.

Campbell, M. D., and J. H. Lehr. 1973. *Water well technology—field principles of exploration drilling and development of ground water and other selected minerals.* New York: McGraw-Hill. 681 p.

Driscoll, F. G. 1986. Groundwater and wells. St. Paul, MN: Johnson Division. 1089 p.

Ferris, J. G., D. B. Knowles, R. H. Brown, and R. W. Stallman. 1962. Theory of aquifer tests. Water-supply paper 1536-E. U.S. Geological Survey. 174 p.

Feth, J. H. 1973. Water facts and figures for planners and managers. Circular 601-I. U.S. Geological Survey. 30 p.

Gibson, V. P., and R. D. Singer. 1977. *Water well manual—a practical guide for locating and constructing wells for individual and small community water supplies.* Berkeley, CA: Premier Press. 156 p.

Heath, R. C. 1983. Basic ground-water hydrology. Water Supply paper 2220. U. S. Geological Survey. 85 p.

Kruseman, G. P., and N. A. de Ridder. 1991. Analysis and evaluation of pumping test data. Publication 47. International Institute for Land Reclamation and Improvement, Wageningen, The Netherlands.

Lohman, S. W. 1972. Ground-water hydraulics. Professional paper 708. U.S. Geological Survey. 70 p.

Public Health Service. 1963. *Manual of individual water supply systems.* Publication no. 24. U. S. Department of Health, Education and Welfare, Public Health Service. Washington, DC: Government Printing Office. 121 p.

Roscoe-Moss Company. 1990. *Handbook of ground water development.* New York: Wiley Interscience. 493 p.

Sorensen, E. F., R. B. Stotlemyer, and D. M. Baker, Jr. 1973. Mineral resources and water requirements for New Mexico mineral industries. Circular 138. New Mexico Bureau of Mines and Mineral Resources. 26 p.

Stone, W. J. 1981. Water for industry in New Mexico's future. *Proceedings,* 26th Annual New Mexico Water Conference. p. 26–77. New Mexico Water Resources Research Institute.

U. S. Environmental Protection Agency. 1976. Manual of water well construction practices. Report EPA-570/9-75-001. U.S. Environmental Protection Agency. 156 p.

CHAPTER 11

Ground-Water Contamination

One of the most important challenges in the world today is cleaning up and protecting ground-water supplies. Studies by hydrogeologists play a critical role in this endeavor. However, such studies do not help resolve such problems unless they are comprehensive and conscientiously conducted. This chapter offers some guidance in the design and implementation of hydrogeologic studies of contaminated ground-water sites. Additional information may also be found in texts by Bedient and others 1994; Domenico and Schwartz 1998; Fetter 1993; and Sanders 1998.

Once existing or threatened contamination is detected, the hydrogeologist must see that the appropriate regulatory agency is notified, the site is characterized, remediation measures are implemented in a timely fashion, and steps are taken to monitor the cleanup. Legal action often follows discovery of ground-water contamination, and hydrogeologists may also be called upon to give expert testimony in court proceedings.

DETECTION

Although both existing and potential ground-water contamination are of concern, the two are detected in different ways. Existing contamination is detected in water supplies or monitoring wells, whereas the potential for ground-water contamination is determined largely through personal observations of discharges or sources of potential contamination. More specifically, people may report accidental spills or suspicious dumping. Similarly, when government employees examine waste-disposal practices and places where the discharge of contaminants occurs or has occurred, the potential for ground-water contamination may be revealed.

Existing Contamination

Contamination may already exist due to a history of human practices that did not protect ground water. Although we commonly associate ground-water contamination with large industrial enterprises (manufacturing, weapons production, mineral extraction, etc.) operating over a long period of time, it may be caused by everyday activities at small businesses and households as well (dry-cleaning establishments, gas stations, jewelry-making shops, etc.). Agricultural activities are also known to produce ground-water contamination (from animal waste, fertilizer, and pesticides). The improper disposal of solid or liquid wastes by municipalities is another source of existing contamination.

The detection of existing contamination in the water supply is different for public and private systems. That is, the process by which contamination of a municipal supply well is discovered is different from that by which a domestic well is found to be polluted.

Contamination of a public supply is usually discovered by analyzing water samples. Routine surveillance of water from a supply well that has a long history of noncontamination may suddenly reveal that one or more constituents exceed the drinking-water standard. When this occurs, another sample is analyzed as soon as possible to determine whether the elevated value is real or due to some handling or laboratory error.

A general deterioration in water quality (taste, color, odor) may be the first sign of contamination of a private well. Follow-up analysis of a sample of water from such a well should reveal the offending constituents. A study of samples from additional wells and a search for potential sources in the area may show whether the poor quality is due to natural causes or contamination.

Existing contamination may also be detected through periodic monitoring of ground-water quality around industrial or waste-disposal sites. In such places, contaminants either have been or may be released at or near the surface. A network of monitoring wells is usually installed at such sites to protect ground-water resources by providing early detection (see chapter 12). Detection of contamination in such wells is less of a surprise than it is in supply wells. Nonetheless, when analysis of a sample from a monitoring well suggests a "hit" for some constituent of concern, the procedure is the same as for supply wells: a confirmation sample is analyzed as soon as possible.

Potential Contamination

The threat of ground-water contamination is associated with leaks, spills, and unprotected discharges at or near the surface. This may involve tanks, impoundments, pipelines, barrels, hoppers, or transport vehicles (for example, trucks and railroad cars). Although spills and leaks are unintentional, other releases that threaten to contaminate ground water are deliberate. While some deliberate discharges are made in ignorance of the environmental consequences and the law, others are made knowingly. Although discharge permits required by state agencies minimize such illegal releases, they nonetheless occur.

Potential ground-water contamination may be detected in various ways. The operators of tank farms, pipelines, trucking firms, railroads, etc., are usually the first to know about and report leaks and spills. Private citizens may also observe and report cases of illegal dumping or discharges. The inspection by government officials of facilities that generate waste also turns up cases of potential ground-water contamination. Such inspections may be conducted in accordance with discharge permits or in response to citizen reports of illegal activity.

Some hydrogeologic studies are conducted to prevent ground-water contamination. Such studies are typically made at potential waste-disposal sites to determine their suitability. Because there is no contamination at such sites, investigation focuses on characterizing the hydrogeologic setting (discussed later in this chapter). This investigation may involve conducting specific tests on selected hydrologic parameters. For example, infiltration or "perc" (percolation) tests may be conducted to determine the rate of water uptake by the soils at the site. Alternatively, a natural tracer may be used to evaluate water movement, thus the contamination potential at waste-disposal sites. Stone (1991) used soil-water chloride as a natural tracer to determine the downward flux of moisture and potential contaminants at various types of waste-disposal sites. The rationale was that sites having a lower flux are more suitable for waste disposal than those characterized by a higher flux.

NOTIFICATION

When involved with a ground-water contamination case, the first thing a hydrogeologist should do is ensure compliance with existing regulations. More specifically, the appropriate regulations must be identified and plans made to meet all requirements within the time intervals prescribed. This may involve state or federal regulations.

In the United States, local (county and state) agencies often had laws and authority for overseeing environmental issues in place before the U.S. Environmental Protection Agency was formed. Some have maintained or regained primacy for regulating certain parts of the environment. Thus, hydrogeologists must be aware of state as well as federal regulations. It is a good idea to become familiar with the organizational units (division, bureau, or section) of the regulatory agency in the state where you work. Each of these agencies may enforce different laws. It is possible to satisfy one set of regulations and be deficient in another, even for the same medium at the same site. For example, it is possible for a site to be given a clean bill of health by the RCRA or UST programs, but still be deficient as far as the state ground-water authority is concerned. Although unlikely, this occurrence is conceivable where metals or petroleum contamination were remediated, but elevated nitrate was not addressed.

The major federal regulations in the United States are summarized in Table 11-1. These are administered by the U.S. Environmental Protection Agency. Usually, a site clearly falls under just one of the federal regulations. Nonetheless, to save time and money later on, it is a good idea to ensure that all pertinent regulatory issues are addressed at the same time. This can only be done if the issues have been reviewed and discussed with the regulators.

The first requirement is usually notification of the state environmental agency that ground-water contamination has been detected or is likely to occur due to an accidental discharge. The onus for notification varies with the situation. For a public water supply, the environmental agency is probably already involved in the reporting of analytical results, if not the sampling, by which contamination was discovered. When contamination is detected in a private water supply, the state agency may be notified directly by

TABLE 11-1 Summary of Regulations Administered by the U.S. Environmental Protection Agency

Act/Amendment	Acronym	Application
Clean Water Act (1972)	CWA	Mainly clean-up and protection of surface-water quality
Safe Drinking Water Act (1974)	SDWA	Public drinking water at the tap
Resource Conservation and Recovery Act (1976)	RCRA	Primarily solid, hazardous materials at treatment, storage, and disposal facilities (except radionuclides)
Hazardous Solid Waste Amendment	HSWA	Reliance on land disposal of untreated waste not acceptable
Underground Storage	UST	Leaking underground storage tanks
Comprehensive Environmental Response, Compensation and Liability Act (1980)	CERCLA (Superfund)	Remediation of past contamination not covered by other laws

the owners of the well or a consultant acting on their behalf. Notification of potential ground-water contamination from leaks or spills may come from a facility owner or employee, a truck driver, the police, the fire department, a government-agency official, or a concerned citizen. Regardless of the situation, if the responsible party has not made such notification, the hydrogeologist should urge that this be done within the timeframe required by law.

Regulations vary from state to state. However, in addition to notification, other requirements generally include characterizing the site, proposing a remediation plan, and installing monitoring wells. Each of these is treated separately next.

SITE CHARACTERIZATION

Obviously, before a site can be cleaned up, it must be characterized. This includes a description of both the natural setting and the contamination. More specifically, a two-part investigation is required: (1) determine the hydrogeologic setting (the local hydrology and its relationship to the regional system), and (2) define the contamination (its nature, concentration, and extent).

Hydrogeologic Setting

The main task for the hydrogeologist in ground-water-contamination studies is the formulation of a sound conceptual hydrogeologic model for the site. There is usually not much time to gather data, but if some previous works on the area can be located, it should not take much effort. Compile basic geologic and hydrologic information (as described in chapters 2 and 5) and characterize the setting (as described in chapters 3 and 6). Identify major hydrogeologic units (see chapter 4) and possible geologic controls of hydrologic behavior (see chapter 7). Then, formulate a conceptual model of the hydrogeology at the site (see chapter 8).

Take care, that, where there are multiple aquifers, water levels and flow directions for all are characterized. This task is important, since the flow direction for shallow perched ground water may be different than that for the deeper regional ground water. Such a difference exists at a site of TCE contamination in an alluvial-fan setting in Albuquerque, New Mexico (Figure 11-1).

Defining Contamination

Once the hydrogeologic setting is established, characterize the contamination itself. Identify the contaminants or constituents of concern, determine the range of concentration for each, and define both the lateral and vertical extent of the plume. In the case of multiple contaminants or discharges, the extent of all plumes must be determined.

Existing and potential ground-water contamination are identified in different ways. For existing contamination, the constituents responsible have no doubt been identified in analysis of ground-water samples. Potential ground-water contaminants, at sites of spills or leaks, may be identified by field methods or lab analysis of soil samples.

The range of concentration of existing and potential contaminants is also determined in different ways. For existing contamination, the range of concentration is readily obtained by examining analytical results for water samples from all nearby monitoring

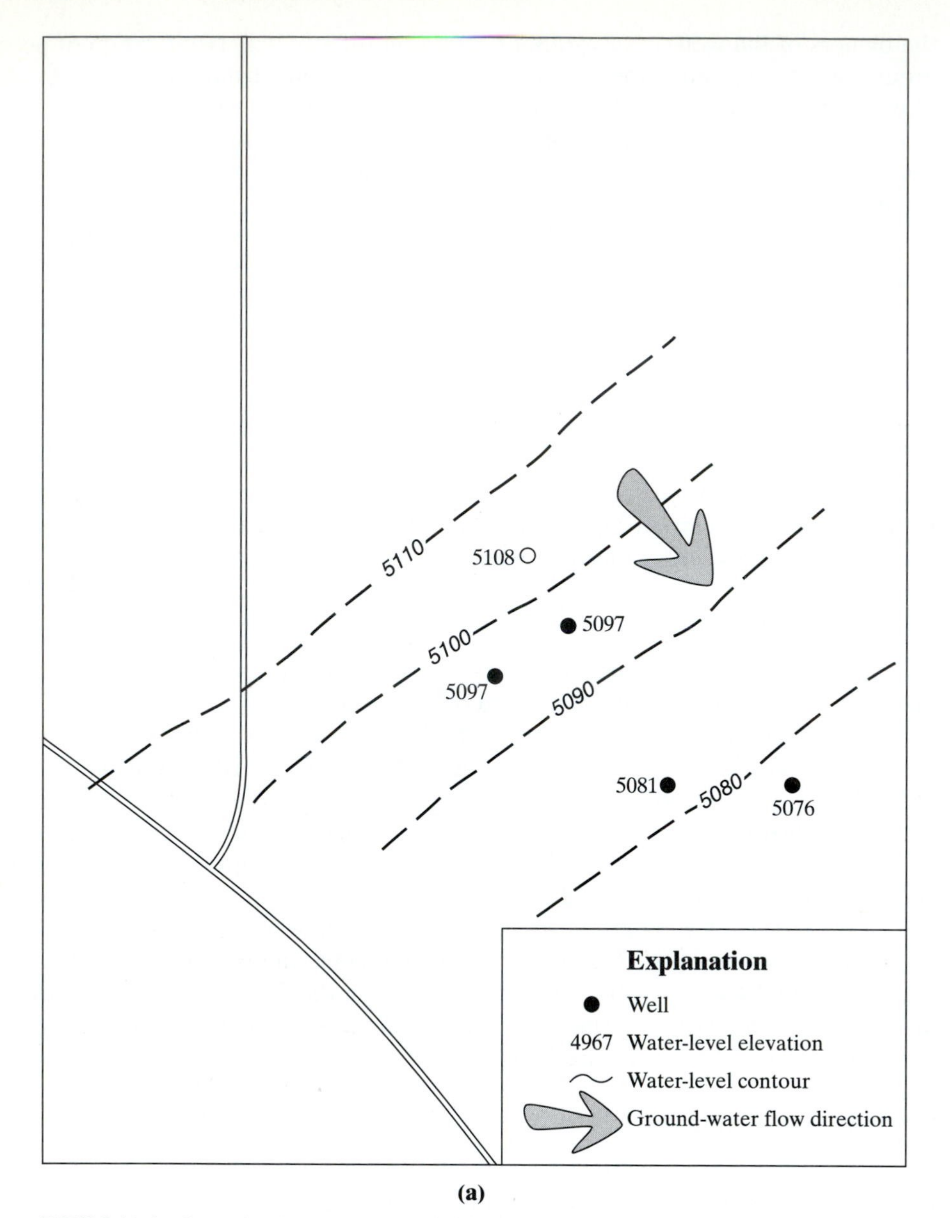

(a)

FIGURE 11-1 Opposing flow directions for (a) shallow perched and (b) deeper regional ground water at a site of TCE contamination, Albuquerque, New Mexico. (Modified from Sandia National Laboratories 1996, Figures 2-5 and 2-6.)

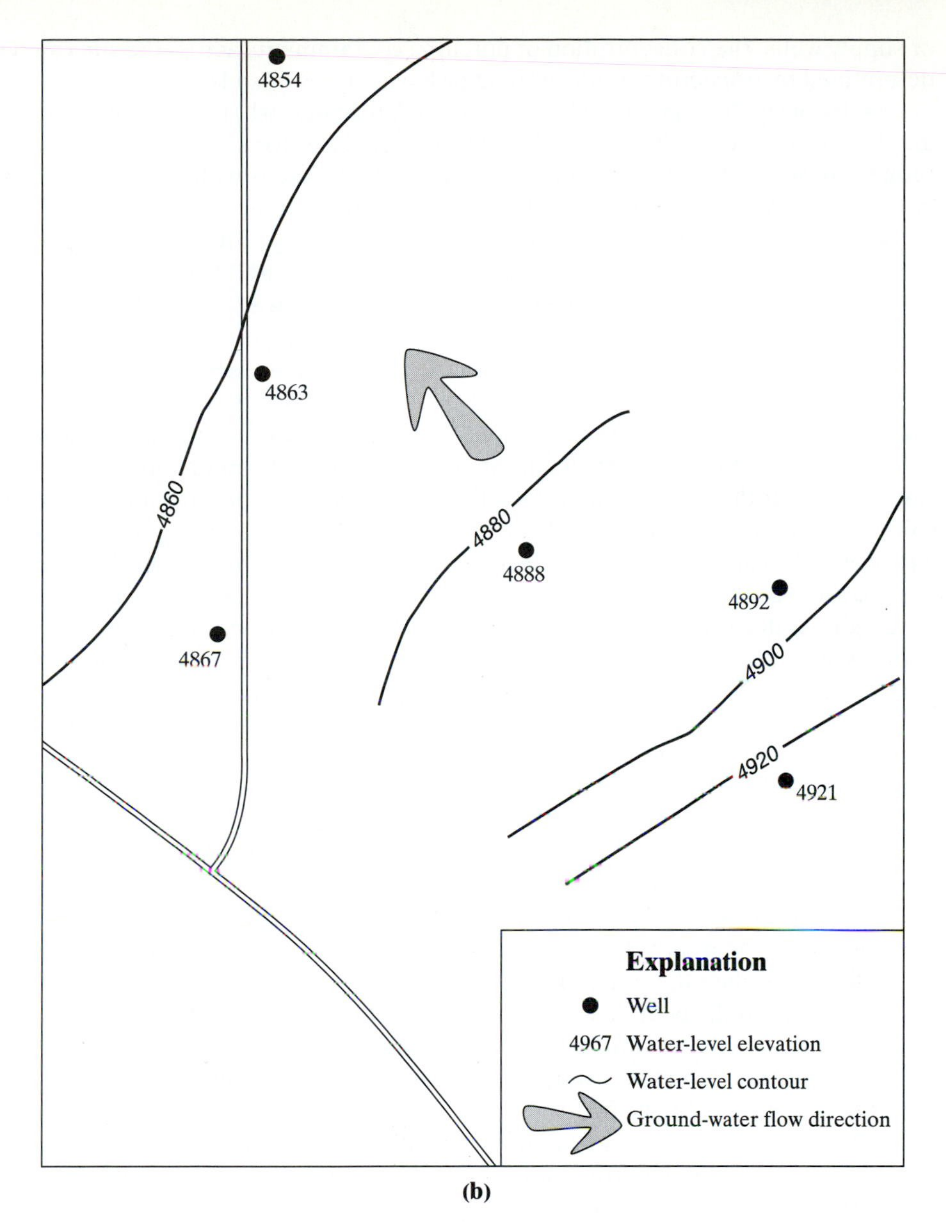
4854
4863
4860
4867
4880
4888
4892
4900
4920
4921
Explanation
Well
4967 Water-level elevation
Water-level contour
Ground-water flow direction

(b)

or supply wells. The concentration of potential contaminants across the site can best be determined by inspecting results of field or lab analyses of soils.

Obviously, the type of analyses specified determines what can be detected. For example, organics cannot be reported by the lab if analyses for such constituents were not requested. Here a bit of homework saves effort later. Take time to determine the history of the site. What was done there dictates what contaminants you should look for. Lab work is expensive, so why analyze for more than necessary? The extra expense and unnecessary delay associated with resampling and making additional analyses for constituents overlooked the first time can be avoided in this way.

In addition to identifying the contaminants at a site, attempt to determine their background concentration in the study area. Obviously, there is no background value for strictly anthropogenic materials (for example, organic solvents, refined petroleum products, etc). However, many constituents of concern occur naturally at some concentration. If background data are not available, analysis of samples from presumably clean locations at or near the site should provide this. Analyses of samples from locations known to be contaminated are not suitable and will probably not be accepted by regulatory agencies. Take care to ensure that the soil types sampled for background concentration are the same as those at the contaminated site. For example, if the soil at the contaminated site is derived from sedimentary rock, samples of soil derived from igneous bedrock may not be appropriate for background analysis because they may naturally contain different concentrations of the constituents of concern.

The lateral extent or edge of the plume is usually considered to be where the concentration of the contaminant corresponds to some acceptable value. This may be the background value for natural materials or the zero contour for anthropogenic materials. Alternatively, a concentration selected relative to a regulatory standard or risk threshold value may be agreed upon by the regulatory agency as marking the edge of the plume.

The method of defining the lateral extent of the contaminant plume is much the same as that for determining the range of concentration. For existing contamination, analytical data for area wells are examined. If there are not enough monitoring points to adequately define the plume extent, additional wells must be installed. Voluntarily adding needed wells is an excellent way to establish good faith with the regulators. Where contaminants are still in the vadose zone, plume definition involves examining results of field and lab analyses of soil samples. For organic contaminants, passive soil-vapor collectors may be useful.

Determining the vertical extent of the plume is done in a similar fashion, but requires analysis of samples from various depths. For existing contamination, data from wells drilled to or screened at different depths are of particular interest. If existing wells do not provide an adequate look at the third dimension, installation of some additional ones may be required. The vertical extent of contamination in the vadose zone can be determined by analyzing soil samples or vapor from different depths. This determination may be accomplished by analyzing soil samples taken by continuous coring with a hollow-stem auger rig or by using driven samplers. Passive collectors are useful if the contaminants are organic.

It is not unusual for ground-water contamination to involve more than one constituent. The earlier this is recognized, the better. Thus, field screening, sampling, and lab analysis should target all suspected contaminants at the outset. Then the range of concentration and extent of the plume for each contaminant detected must be determined.

There is always concern over the quality of analytical data. Two things can be done to minimize such concern. First, obtain a copy of and review the quality assurance/quality control (QA/QC) program used by any analytical laboratory to which you plan to submit samples. If it is not clear to you, show the program to a chemist to see if it conforms to standard practice. Also, once you are comfortable with the lab's QA/QC program, have one of their representatives present it to the regulatory agency that will receive your data. Second, if the state agency you are dealing with has prepared a document on data quality assurance (e.g., like that by the New Mexico Health Department 1992), obtain a copy and follow it to the letter. If you have used their protocol and the lab you use has a sound QA/QC program, acceptance of your data is assured.

REMEDIATION

Once the hydrogeologic setting has been conceptualized and the nature and extent of the contamination defined, remediation approaches may be addressed. Selection of a cleanup method involves weighing three things: protecting the health of both cleanup workers and the public, technical feasibility, and economic considerations.

The specific identity, form, or concentration of the contaminant may pose a serious health threat not only to workers involved in the cleanup but also to the public at large. Such considerations may limit the remediation options available. For example, it may be safer to deal with a potentially inhaled substance in-place than to excavate it, thereby allowing for its broader dissemination by wind, or hauling it through town in open trucks, increasing human exposure.

Even though a given remediation method is acceptable from a health standpoint, it may be unacceptable if it is difficult, ineffective, or creates still further contaminants. Thus, other important considerations in evaluating a possible remediation method are whether it is technically feasible and whether it is likely to produce the desired results in the time required. For example, artificial wetlands have been proven effective in removing metals from waste streams (Figure 11-2). However, if a large volume of waste is involved, a sizeable expanse of wetlands and rigorous rotation of areas in the treatment process may be required. Furthermore, once the plants take up and concentrate the metals, they themselves become a hazardous waste that must be disposed of properly. Such a complex remedy may be too unwieldy for some sites.

Finally, a given remediation option may be acceptable on the basis of both health and technical considerations, but the cost may be prohibitive. For example, in the case of wetlands, the cost of testing any wildlife or livestock feeding on the vegetation may be prohibitive. Also, consider the reverse-osmosis option. Although it poses no health threat to remediation workers or the general public, and it can remove many ions from saline water, the cost is too great for most applications.

In certain circumstances, a single consideration may override all others. That is, a more costly remediation technique may be chosen if it is necessitated by health concerns or if it is the most efficient way to meet target deadlines. For example, the diesel-

FIGURE 11-2 Experimental wetlands operated by the Colorado School of Mines at the Argo Tunnel, Idaho Falls, Colorado. Photo by Thomas Wildeman. Used by permission.

contaminated sediments at the AT&SF railroad yard in Belen, New Mexico, were of such low permeability that pumping out the floating product would have taken too long. Thus, multiple recovery wells were installed in a collection trench, specially dug and filled with permeable material for the purpose (Figure 11-3).

Conversely, a relatively inefficient clean-up method may be selected if it is the both the safest and most economical.

MONITORING

Monitoring is treated in greater detail in chapter 12, but is briefly addressed here to complete the discussion of remediation. As used here, the term applies mainly to the methodical search for pollution. Nonetheless, the flow system at the site can only be evaluated if monitoring includes water-level observations as well.

Monitoring is an important part of ground-water contamination investigations and provides information at all stages of such projects. For example, it may be the means by which the contamination was detected in the first place. During site characterization, analysis of samples from monitoring wells yields not only the identity of contaminants, but also reveals the concentrations necessary to define the lateral and vertical extent of the associated plume(s). During remediation, monitoring provides a means of tracking the progress of cleanup. For example, progress may be shown by the decrease in concentration of a constituent of concern or, where ground water has been contaminated by petroleum, a decline in the thickness of floating product. Monitoring may also be the means by which remediation is judged to be completed. For example, some predeter-

FIGURE 11-3 Wells installed in specially constructed trench to enhance recovery of diesel fuel floating on the shallow water table, AT&SF railroad yard, Belen, New Mexico.

mined number of consecutive "clean" samples, such as four quarters, may be the basis for closing of a contaminated site.

The monitoring tasks of the hydrogeologist vary, depending on their role in the contamination case. If the hydrogeologist is the consultant to the responsible party, he designs, installs, and operates, the observation network, as well as prepares regular reports on monitoring results (monthly, quarterly, or annually). If she is the regulator responsible for the case, she not only reviews the adequacy of the monitoring-network design, but oversees its installation and operation. Additionally, the hydrogeologist evaluates the remediation plan, reviews regular monitoring reports, and sometimes cosamples wells with the responsible party's consultant. In reviewing monitoring reports, the hydrogeologist pays particular attention to any changes in parameters that may indicate a spread of the contaminant plume or the effectiveness of the remediation process.

EXPERT TESTIMONY

When contamination is in serious violation of state or federal law, remediation efforts are deemed by the regulator to be deficient, or the responsible party defies regulatory orders, legal action may result. The hydrogeologist involved with the case is commonly asked to appear in court to provide expert testimony on the site.

There are various sources of guidance on performing this task. An extensive treatise on the subject is that by Bradley (1983). A very useful booklet on being an expert witness has been prepared by the American Institute of Professional Geologists, (AIPG) (1994). A paper by Wyche (1995) gives a lawyer's perspective on the topic. Although some of the more important considerations are listed next, the interested reader is referred to those sources for further details.

Appropriate courtroom behavior is largely a matter of common sense. Nonetheless, a few basic guidelines from the AIPG booklet and observations from personal experience should be instructive:

1. To preserve the appearance of impartiality, do not sit with representatives of the side you were retained or called by.
2. Together with the attorney, formulate the nature and order of key questions to be asked, and rehearse your testimony just before the trial.
3. Don't try to impress the court with technical jargon; your input has little value if it cannot be understood. Keep testimony at the layperson's level.
4. Make any graphics presented simple and easy to read from the distance required.
5. Do not be rushed into hasty answers; take your time.
6. You may refer to notes to refresh your memory, but be prepared to surrender them for examination by the opposing attorney.
7. Remember, an expert is the only witness that is allowed to voice an opinion in a court proceeding. When asked for an opinion, base it on sound principles and offer it confidently.
8. Do not guess; if you do not know the answer to something, say so.
9. Feel free to admit a mistake or qualify an answer.
10. In short, be honest, ethical, and professional.

You may also be asked to suggest questions to be asked of the other side's expert witness. If so, consider this carefully and keep them central to the issues in the case.

Integrity is essential, regardless of whether you represent a government agency, the perpetrator, or an injured party. If you are a government hydrogeologist, it is imperative that you are thoroughly familiar with your agency's regulations and administer them equally to all. If you are a consulting hydrogeologist, it is a good policy to establish with a client at the outset that you will report what you find, not what they want to hear. This will avoid any embarrassment later on. Protect your professional reputation by disassociating yourself from any client involved with fraudulent activities or unscrupulous lawyers.

Ground-water contamination studies are probably the most important tasks that a hydrogeologist does for several reasons. First, public health depends on the outcome. Second, the responsible party may incur considerable costs in monitoring and remediation. Third, such studies are likely to come under more scrutiny than any other, often receiving media coverage. Thus, the reputation of your employer is at stake. Regardless of who you are working for, it is imperative that you do the most thoughtful and thorough study possible, given the conditions surrounding the case.

In response to the increase in ground-water contamination cases and projects, a number of recent texts have been devoted to the subject (Bedient and others 1994; Domenico and Schwartz 1998; and Fetter 1993). Furthermore, various journals are now devoted exclusively to the subject (e.g., *Ground Water Monitoring and Remediation* and *Journal of Contaminant Hydrogeology*). Consult these sources for further information; for additional case histories, as well as for new developments in monitoring and remediation technology.

CASE HISTORY

Since most incidents of ground-water contamination involve hydrocarbon or organic materials, the case selected deals with petroleum. Although each case is unique in some way, the approach to the hydrogeologic studies they prompt is similar. The following example is typical.

Pipeline Leak, Endee, New Mexico

Early one Sunday morning a major high-pressure petroleum pipeline between Amarillo, Texas, and Albuquerque, New Mexico, sprang a leak, spraying jet fuel far up into the air. In response to a southeasterly wind flow, the spray fell on the ground surface over an elongate area having the approximate dimensions of 600 ft long and 15 to 150 ft wide (widest at the source of the leak). An estimated 2,088 barrels (bbls) (87,695 gal) were discharged before valves straddling the leak were shut off and flow was stopped. The spill damaged some 17,000 yds^3 of soil used for agricultural purposes and threatened the ground water tapped by wells at two adjacent ranch houses.

Detection

The pipeline company was the first to discover the leak. It was located during a search for the cause of a pressure drop in the line, detected in Amarillo. Had this not occurred, the leak would have eventually been detected on-site, as the impressive fountain of fuel that developed was clearly visible (and audible) from the road and nearby ranch house.

Notification

This case fell under the jurisdiction of the state environmental agency (New Mexico Environment Department). More specifically, Article 1-203 (1203 in the December 1995 revision) of the New Mexico Water Quality Control Commission's regulations applied (1995). That article covers

> "... discharge from any facility of oil or other water contaminant, in such quantity as may with reasonable probability, injure or be detrimental to human health, animal or plant life, or property, or unreasonably interfere with the public welfare or the use of property...."

The regulations specify that the Chief of the Ground Water Quality Bureau of the New Mexico Environment Department be notified as soon as possible after detection or within 24 hours. Because the leak occurred on a weekend, the notification was made through the agency's emergency hot line. The bureau that operates the hot line forwarded it to the Ground Water Quality Bureau.

Site Characterization

A county ground-water report, available from the state geological survey (Berkstresser and Mourant 1966), provided both the pipeline operator's consultant and the regulator with geologic and hydrologic coverage of the site, at least at a regional scale. Geologically, the area was found to be characterized by Triassic redbeds (Chinle Formation) overlain by Quaternary eolian sand. The redbeds generally have low permeability and regional water table lies at shallow depth within them. Regional ground-water flow is northeasterly.

The leak occurred at the top of a slope covered by eolian sand. Wind blowing at the time spread the petroleum over an oval area. The fuel-saturated sand clearly marked the lateral extent of the surface fallout from the leak. Field and lab measurements verified the visual edge of the contaminated soil.

To determine the vertical extent of contamination, six monitoring wells were installed. These wells revealed several things:

1. Unconfined water occurs at shallow depths, ranging from approximately 16 to 36 ft, where it may be perched atop a well-cemented zone within the redbeds.
2. Flow direction for this water is southeasterly, in response to local topography (Figure 11-4).

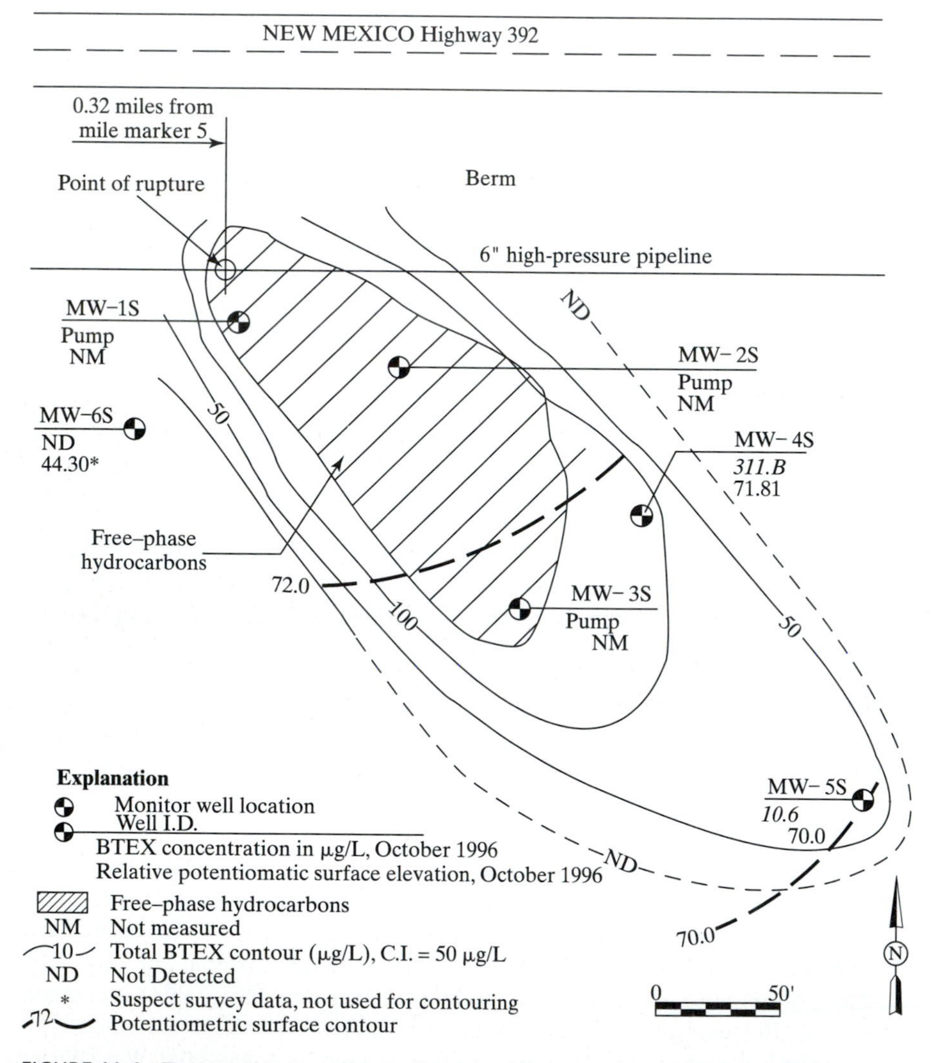

FIGURE 11-4 Extent of plume and ground-water configuration at the pipeline leak site, Endee, New Mexico. (Modified from Geoscience Consultants Limited 1996, Figure 1.)

3. Up to 8 ft of free product was found to be floating on this perched water.
4. Concentration of major fuel components (benzene, toluene, ethylbenzene, and xylene or BTEX) outside the area of free product ranged from 0 to >100 μg/L.
5. The extent of ground-water contamination was similar to that of the surface contamination (Figure 11-4).

No organic contaminants (BTEX) were detected in periodic analysis of water from the ranch-house wells.

Remediation

A remediation plan was submitted to the state by the pipeline operator through its consultant. It called for recovering free product by means of pumping three of the six wells installed. This was stored in a tank on site (Figure 11-5); when the tank was full it was emptied by truck and taken to the pipeline company's nearby terminal for oil and water separation. Quarterly reports gave the amount of product recovered (Table 11-2). Clean-up of the contaminated soil was left to natural bioremediation.

Runoff from a heavy rain storm carried fuel-contaminated water and sediment down slope (Figure 11-6). In fact, it flowed as far as the vicinity of the closest ranch house. To prevent this from happening again, earthen berms and fabric silt dikes were placed on the slope to divert and retard overland flow of storm runoff (Figure 11-7).

FIGURE 11-5 Recovery system and storage tank, pipeline leak site, Endee, New Mexico.

TABLE 11-2 Example of Hydrocarbon Recovery Data

Date	Comment	Cumulative Product Recovered (gal)
1-17-94	Removed 500 gallons from tank	2,650
2-7-94	Installed pump in MW-3S	
2-7-94	Removed 250 gallons from tank	2,900
3-2-94	Removed 200 gallons product; 200 gallons water from tank	3,100
3-15-94	Removed 80 gallons product; 170 gallons water from tank	3,180
3-29-94	Removed 120 gallons product; 100 gallons water from tank	3,300
4-18-94	Removed 225 gallons product; No water	3,525
5-2-94	Removed 160 gallons product; No water	3,685
6-7-94	Removed 200 gallons product; 300 gallons water from tank	3,885
6-29-94	Removed 250 gallons product; 250 gallons water from tank	4,135
July	No product removed	
August	No product removed	
9-22-94	Removed 300 gallons product: 100 gallons water from tank	4,435
10-3-94	Removed 350 gallons product; 25 gallons water from tank	4,785
11-16-94	Adjusted pump levels	
11-24-98	4 1/2" in tank	
11-26-94	7" in tank	
12-20-94	Removed 250 gallons product; 100 gallons water from tank	5,035

Source: Modified from Geoscience Consultants Limited, 1996.

FIGURE 11-6 Evidence of scour by storm runoff, pipeline leak site, Endee, New Mexico. The lower half of the boards around the concrete pad was originally buried in the eolian sand at the surface here.

FIGURE 11-7 Silt curtains installed to minimize erosion of contaminated soil by storm runoff at the pipeline leak site, Endee, New Mexico.

Monitoring

The six wells installed provided information about the local hydrogeology and the distribution of contamination. Quarterly sampling and analysis of ground water for the comprehensive suite of organic constituents associated with jet fuel provided a means of checking the progress of recovery as well as the migration of floating product and dissolved constituents.

Discussion

Note that the leak was shut off in a remarkably short time after it occurred, considering it happened early in the morning on a weekend. Contamination would have been much worse had the leak been allowed to flow longer. A pin-hole leak probably existed in the pipeline for some time before the rupture and associated fountain finally developed.

As hydrogeologist for the regulatory agency, I worked closely with the rancher, pipeline company, and its consultants. An employee based in our field office near the leak alerted me to conditions and significant activity at the site, between my visits. This was invaluable as I was based in another office, more than 200 miles away.

Although this case involved a petroleum spill, ground water may also be contaminated with metals, radionuclides, organic solvents, or pesticides. Thus, many other case histories may be found in the literature and among state-agency files. These histories can be of considerable help to those new to the ground-water contamination field. Regardless of whether you are a consultant or a regulator, examples of work on ground-water contamination cases similar to the one at hand should be reviewed before conducting a study. Examples of previous work by your employer are especially useful, since they can provide the in-house preferences regarding such things as field methods, lab analyses, data forms, report format, etc.

REFERENCES

American Institute of Professional Geologists. 1994. *The professional geologist as expert witness.* Monograph 4. American Institute of Professional Geologists. 20 p.

Bedient, P., H. Rifai, and C. Newell. 1994. *Ground water contamination, transport and remediation.* Upper Saddle River, NJ: Prentice-Hall. 541 p.

Berkstresser, C. F., Jr., and W. A. Mourant. 1966. Ground-water resources and geology of Quay County, New Mexico. Ground-Water Report 9. New Mexico Bureau of Mines and Mineral Resources. 115 p.

Bradley, M. D. 1983. The scientist and engineer in court. Water Resources Monograph 8. American Geophysical Union. 111 p.

Domenico, P., and F. Schwartz. 1998. *Physical and chemical hydrogeology.* New York: John Wiley & Sons. 506 p.

Fetter, C. W. 1993. *Contaminant hydrogeology.* New York: Macmillan. 458 p.

Geoscience Consultants Limited. 1996. Fourth Quarter Report for the Diamond Shamrock Endee Site, December 1996. Consulting letter report. Geoscience Consultants, Limited.

New Mexico Health Department. Surface Water Quality Bureau. 1992. Quality assurance project plan for water quality management programs. New Mexico Environment Department and Scientific Laboratory Division, New Mexico Health Department.

New Mexico Water Quality Control Commission. 1995. New Mexico Water Quality Control Commission Regulations. New Mexico Water Quality Control Commission. Santa Fe, NM.

Sanders, L. L. 1998. *A manual of field hydrogeology.* Upper Saddle River, NJ: Prentice-Hall. 381 p.

Sandia National Laboratories. 1996. Sandia North groundwater investigation plan. Sandia National Laboratories. Environmental Restoration Project.

Stone, W. J. 1991. Estimating contamination potential at waste-disposal sites using a natural tracer. *Environmental Geology and Water Science.* 19(3): 139–45.

Wyche, B. W. 1995. The hydrogeologist in litigation—top ten tips on how to be an effective expert witness. *Ground Water Monitoring and Remediation* 15 (2): 94–96.

CHAPTER 12

Hydrologic Monitoring

Monitoring may be defined as observing some parameter or condition over a period of time. Such surveillance is not new to the field of hydroscience. Precipitation, stream flow, and river quality have been monitored for some time. Similarly, ground-water level and quality have long been monitored, especially in public water-supply studies. For example, in 1968, nearly 29,000 water-level observation wells were in use in the United States alone (Pauszek 1972). Water-level drawdown is also routinely monitored in wells during pumping tests. However, with the growth of the environmental industry, the scope of monitoring has expanded. Monitoring one or more aspects of the hydrologic system is essential in environmental studies. In some cases, such surveillance is required by law.

A single chapter can provide little more than an overview of the main issues in monitoring and the interested reader is directed to the large body of literature that has been generated on the subject. A separate journal, *Ground Water Monitoring Review,* was established in 1981 as an outlet for basic information on the growing field of hydrologic monitoring (the journal was renamed *Ground Water Monitoring and Remediation* with volume 13, Winter 1993). Volume 1, Number 1 of that journal gave a series of papers covering major aspects of monitoring from theoretical considerations to well construction. Much has been learned since then, as subsequent issues of this periodical demonstrate. However, those initial papers provide a good introduction to the topic for newcomers. The article by Kazmann (1981) is an excellent overview of the problems and processes of ground-water monitoring.

Procedures for monitoring specific media or hydrologic parameters may be found in some general sources, as well as more specific references. Those new to the field should especially consult *The National Handbook of Recommended Methods for Water-Data Acquisition* (U. S. Geological Survey 1977) and the series on *Techniques of Water-Resource Investigations of the United States Geological Survey.* These are excellent resources to use when setting up a monitoring system. Booklets on basic surface-water measurements are included in the "techniques" series (e.g., those on stream gaging by Carter and Davidian 1968; and Buchanan and Somers 1968). Entire books are devoted to soil-water or vadose-zone monitoring (e.g., those by Everett and others 1984; and Nielsen and Johnson 1990). The comprehensive guide to constructing monitoring wells, available from the U.S. Environmental Protection Agency (Aller and others 1991), is especially useful in ground-water surveillance projects.

Hydrogeologists are involved at various stages in monitoring: design, installation, operation, and evaluation. The purpose of this chapter is to examine each of these stages, emphasizing the various tasks involved and decisions required, then to show how these tasks are applied in practice by reviewing some case histories.

DESIGNING A MONITORING SYSTEM

Setting up a monitoring system requires careful consideration of both the hydrogeologic setting and the data needed. Using this information, you should be able to design an appropriate monitoring plan.

Hydrogeologic Setting

The hydrogeologic setting is the framework within which observations are carried out. Thus, monitoring must be based on a sound conceptual hydrogeologic model (as discussed in parts I and II and summarized in chapter 8). Only then will the number and location of monitoring stations be appropriate and the monitoring methods be adequate. As noted by Summers (1985),

> The design of a monitoring system depends upon the designer's perception of how the environmental factors merge with his[her] perception of the ground-water flow system. The validity and representativeness of measurements made during a monitoring program depend upon the validity of the designer's perceptions and concepts. To optimize the design of a ground-water monitoring program the designer should make preliminary measurements to confirm his[her] assumptions.

The best way to confirm assumptions about the ground-water flow system in an area is to compile water-level information. Water-level information is not only needed prior to designing the system, but during the monitoring period. Some of the wells in monitoring networks are often dedicated to this task alone. Suggestions for the design of observation-well programs were given by Heath (1976); the objectives and products of the major types of water-level networks he recognized are summarized in Table 12-1.

Although water-quality information may not be available until monitoring wells are installed, samples from existing wells should be analyzed. As in the water level investigation, a preliminary understanding of area ground-water quality is also useful in designing the monitoring system. The type of network established depends on the purpose of the water-quality monitoring (Table 12-2).

TABLE 12-1 Types of Water-Level Monitoring Networks

Network Type	Purpose	Results
Hydrogeologic	Extent of aquifers/storage	Regional water-level maps
	Hydraulic properties	Hydrographs: water level versus time
	Degree of confinement	Graphs: water level versus pumping rate
Water-resource management	Baseline conditions	Prepumping water-level maps
	Effect of stress on recharge/discharge	Local water-level maps
	Status of storage	Water-level change or storage-change maps

Source: Modified from Heath 1976.

TABLE 12-2 Types of Ground-Water–Quality Monitoring Networks

Network Type	Purpose	Results
Regional	Water-resource quality Hydrochemical facies Natural concentrations for specific constituents	Regional water-quality maps Hydrochemical facies maps Background water-quality–values, maps
Site-specific	Site characterization Surveillance Post-remediation contaminant levels	Plume maps Periodic water-quality maps Verification of cleanup

Data Needed

The data needed vary from study to study and site to site. However, in all cases, designing a system to obtain data requires a determination of what might be called the monitoring "target," "focus," and "objective." The target is the hydrologic medium to be monitored, the focus is the parameter(s) of interest, and the objective is the reason for the surveillance.

The target may include surface water, soil water, ground water, or some combination of these. The target, together with the focus and objective, determines the type of monitoring equipment and procedures to be employed. These in turn, dictate how long it will take and how much it will cost to obtain the data required.

The focus of monitoring may be water quantity, water quality, or both. Water quantity issues include such factors as the flow of a major stream, the variation in selected parameters of the local water budget, or water-level decline in a ground-water basin. Water-quality concerns might include general surface-water parameters (pH, temperature, suspended load, etc.), the chemistry of storm runoff, the distribution of a given contaminant in soil at a spill site, or the amount of various constituents in the water supply.

Usually the objective of monitoring is to answer a specific question or set of questions. What is the ground-water flow direction? What is the background concentration of a constituent of concern? What contaminants are present? What is the extent of a contaminant plume? How is a plume migrating? Successful monitoring is not so much a matter of getting the right answer as it is asking the right questions. Thus, the objective should be carefully formulated.

The objective also determines the scale and timing of the surveillance. Is it to be regional or site-specific? Is it to be conducted at regular intervals or continuously? Is it to be short- or long-term? Such issues should be carefully considered in defining the objective.

Monitoring Plan

Once the hydrogeology of the site is conceptualized and the data needs are established, the monitoring plan can be designed. This includes selecting the equipment and methods to be used, determining the number and location of monitoring stations needed, and deciding on the frequency of both taking and reporting measurements. These decisions must consider not only the target, focus, and objective, but also site conditions, budget constraints, and staff capabilities.

FIGURE 12-1 Snow gage in Tuscarora Mountains, Carlin Gold Trend, northern Nevada.

Selection of monitoring equipment and methods depends, in part, on site conditions, especially climate. For example, if year-round precipitation measurements are needed, it does little good to use unheated recording rain gages in an area where they will freeze up four months of the year. It is more reasonable to use heated recording gages, where a reliable power supply is available, or nonrecording all-weather rain and snow gages, where it is not (Figure 12-1). Similarly, stream gages that record stage every 5 minutes are wasted on ephemeral streams in an arid setting.

Equipment and method decisions are also driven by costs. Can the information needed be obtained without the latest, cutting-edge gizmo? In many cases, the low-tech approach is at least as good and sometimes better. For example, a more cost-effective means of monitoring ephemeral stream flow is to use weirs or flumes and crest gages that can be visited immediately after a runoff event is likely, based on the local weather conditions (Figure 12-2). Using dedicated pumps in ground-water monitoring programs may cost more up front, but be more cost-effective in the long run because the expenses associated with decontamination, as well as resampling and analysis due to cross-contamination of samples, are eliminated.

FIGURE 12-2 Weir on ephemeral stream, Carlin Gold Trend, northern Nevada.

Any monitoring equipment or methods selected must also be appropriate for the staff that will be responsible for them. It makes no sense to employ equipment or methods that are beyond the technical ability of the available staff. This is especially true if training is not an option. In some projects, however, it may be necessary to provide training on new equipment or methods. This may be accomplished through a commercially available professional short course or customized training, arranged with the equipment manufacturer, the USGS, or others familiar with that type of monitoring.

The number of monitoring stations depends to a large extent on the size and hydrogeologic variability of the study area. Just because an area is large doesn't mean it must be riddled with monitoring equipment. A related concern is that the more stations there are, the more analyses there will be. What is the smallest number of stations that will provide basic coverage? For example, if springs occur in widely spaced clusters, is it necessary to monitor every spring in each cluster? In such a case, preliminary monitoring should include all of the springs. If results show little variation between springs, each cluster may be represented by a single station.

The location of observation points depends both on the number decided upon and what is known of the study area. Where are the possible sources of contamination? What is the conceptual hydrogeologic model for the area? These provide guidance on where to place stream gages, runoff samplers, neutron-probe access tubes, monitoring wells, etc., in order to encounter any contaminants.

Take care to not only achieve a reasonable number of stations, but also a reasonable suite of analyses required for samples from each station. Since sample analyses can be expensive, they should be kept to a reasonable level. What is the minimum number and type of analyses that must be run in order to meet the monitoring objective? Note that the term *monitoring requirement* (as defined by regulations) is not used. Too often the minimum becomes the maximum. For example, if there is a regulation that requires septic tanks to be at least 5 ft from the house, then all septic tanks tend to be exactly 5 ft from the house, even where there is room to place them at a greater distance. If a minimum is suggested, whether for a distance or a suite of analytes, people tend to take it as a hard and fast specification. The point is that the minimum doesn't necessarily fit all situations; if more stations, sampling, analyses, etc. are required for better site characterization, more adequate monitoring, or just good science, they should be included.

Staff time is also expensive and may be a limiting factor. How many people are needed to cover a given monitoring network? Alternatively, how long will it take available staff members to cover an optimal network? One solution to staff constraints is sharing monitoring duties with another group, who may be another responsible party or a government agency. In such a scheme, each party covers a specific portion of the surveillance network, but the portions overlap enough to provide a means of checking results.

Frequency of observations is a critical item in the monitoring plan. Generally speaking, measurements are made more frequently at the outset of monitoring and less so once seasonal variation has been established. Monthly and even weekly measurements may be justified at the beginning of monitoring; quarterly observations may be sufficient later on. Stream monitoring may eventually be limited to peak runoff seasons. Similarly, water-level monitoring may be tailored to the known or suspected recharge and nonrecharge seasons. In irrigated areas, water levels are commonly measured between growing seasons, when they are least likely to be stressed, say in late winter. For example, the USGS monitors water levels in New Mexico for the state engineer by taking annual readings in February (as long after pumping has ceased as possible, but before the next period of irrigation begins).

What should be the frequency for reporting monitoring results? Quarterly is a common frequency for both observations and reporting. However, the frequency of reporting need not be the same as that for monitoring. For example, observations may be made quarterly but reported annually. Some observations may be continuously recorded (such as stream flow or water level) and others taken irregularly (such as storm runoff). If numerous water samples are involved, the time required for their analyses will dictate when a report can be prepared.

What form should data reporting take? Unless regulations specify otherwise, results of chemical analyses, water-level measurement, etc., may be conveyed informally. This may simply consist of a map, showing the location of monitoring stations, accompanied by a table of data. Such tables should include previous results for comparison (see Table 12-3). Ideally, however, a report is prepared that not only presents the monitoring data, but interprets them relative to health risks, remediation progress, etc.

INSTALLING A MONITORING SYSTEM

Once a hydrologic monitoring system has been designed, it must be installed. This involves emplacing the surface-water, soil-water, and ground-water monitoring stations prescribed at the sites selected.

TABLE 12-3 Comparison of Historical Data for Thickness of Floating Product in Some Monitoring Wells, Diesel Spill, AT&SF Railroad Yard, Belen, New Mexico.

Well I.D.	3/93	6/93	10/93	12/93	3/94	6/94	9/94	12/94	3/95	6/95	9/95
MW-5	0	0	0	0.02	0.12	0	0	0	0	0	0.02
MW-8	NM	NM	NM	3.61	4.27	2.73	4.24	3.57	3.43	4.07	4.14
MW-15	3.24	2.52	3.31	3.42	3.53	3.45	2.57	2.75	2.14	3.66	3.32
MW-18	0.67	0.72	0.05	0.40	0.52	0.91	0.55	0.76	1.40	0.95	1.13
MW-19	1.37	0.96	0.93	NM	1.55	0.57	0.69	0.48	0.51	0.47	0.53
MW-22R	3.62	3.55	3.36	3.49	3.61	3.27	3.03	3.37	3.23	3.48	3.54
MW-23	>2.14	>2.84	>3.15	>1.70	>1.48	>2.21	3.01	1.79	2.18	2.08	2.06
MW-30R	5.07	5.45	5.56	5.30	5.09	4.93	4.87	5.13	4.64	5.29	4.90
MW-37	0	0	0	0	0	0	0	0.87	0.66	2.06	2.47
MW-38	0.68	0.60	0.63	0.91	0.99	0.52	0.38	0.68	0.35	0.89	0.89
MW-40	0	0	0	0	0.01	0	0	0	0	0	0
MW-41	0	NM	NM	NM	NM	0	0	<0.01	0	0	0
MW-43	3.06	3.08	3.14	3.05	2.86	2.95	3.36	3.23	3.27	3.41	3.32
MW-44	0.88	0.68	0.72	1.40	1.57	0.62	0.73	0.90	0.54	1.19	1.26
MW-49	5.86	6.07	5.86	5.60	5.35	5.82	5.50	5.79	5.18	5.24	5.13
MW-51	NA	NA	NA	NA	NA	0.02	0.02	0.03	0.02	0.04	0.09
MW-54	<0.01	<0.01	0.46	0.75	1.73	4.28	5.52	5.05	4.02	4.47	4.63
MW-56	5.15	6.03	6.15	5.60	5.17	5.99	6.21	5.83	5.85	5.59	5.71
MW-59	3.53	4.17	4.06	3.35	3.16	3.85	4.61	3.69	3.54	3.54	3.80
MW-60	0.20	0	0.61	0.90	0.58	0.76	0.15	0.65	0.27	1.04	0.77
MW-65	NA	NA	NA	NA	NA	5.54	5.92	5.48	5.70	5.78	5.54
MW-66	NA	NA	NA	NA	NA	6.34	6.84	6.12	5.53	5.34	5.63
MP-1	1.92	1.55	2.14	2.35	2.69	1.65	1.41	1.37	1.32	1.64	1.58
MP-2	1.86	NM	1.84	2.46	>1.36	1.32	0.87	0.83	0.60	1.02	0.74
MP-3	>1.61	2.03	1.83	1.52	1.41	1.14	0.64	0.85	0.36	0.91	0.69

Source: From Radian Corporation 1995.

Surface Water

Although general locations of weirs or flumes, recorders, and samplers may have been determined, site-specific conditions will dictate their ultimate placement. Although sites affording continuous contact with flow should be selected, it is not always possible. Consider the Rio Salado, a wide, braided, ephemeral tributary of the Rio Grande in central New Mexico. For many years the USGS operated three gages located across its width. The Salado often flowed without getting any of the gages wet! Narrow reaches with distinct banks and a hard or bedrock channel bottom not only provide good control of the flow, but also permit anchoring of weirs or flumes. In arid settings, a solid bottom is often provided by a layer of ground-water caliche or calcrete (calcite-cemented sand and gravel). Such control is supplemented by sand bags or concrete, if necessary. Ideally, sampling is done at the same place where flow is measured so the volume of constituents of concern may be calculated. Another major consideration, of course, is whether the site can be reached with the heavy equipment necessary to install the station (back hoe, drilling rig, etc.). Also, if monitoring equipment is to be solar-powered, does the site receive sunlight year-round?

Soil Water

Installing specific equipment to determine soil-water properties and chemistry may be required if the vadose-zone is to be monitored. As for the installation of surface-water stations, site-specific conditions must be considered, especially accessibility and suitability for using solar power, if anticipated. Characterizing hydraulic properties of the unsaturated material may include installation of such items as neutron-probe access tubes and tensiometers. Preliminary work may be required to determine the suitability of various methods. For example, if porous-cup lysimeters are to be used, is suction within the operable range of such devices? If not, monitoring soil-water chemistry may require installation of other devices, such as active soil-vapor probes or passive soil-vapor samplers. Take care during installation of soil-water monitoring equipment that nothing is introduced that will bias the results.

Ground Water

Monitoring ground water requires drilling and constructing monitoring wells, as well as setting up measuring and recording equipment. As when installing surface- and, soil-water monitoring equipment, well locations should be evaluated for accessibility and suitability for solar power. Monitoring wells are an important part of most surveillance networks and will be discussed in greater detail later in this chapter.

Safeguarding Equipment

Various expensive and sensitive equipment may be set up as part of a monitoring network. Some require more or less permanent installation. Take great care to ensure that they are not damaged and the record is not interrupted once monitoring begins. Thus, it is important to protect recorders from washout, weather, and vandals. This may involve building special platforms, constructing support structures, installing protective barriers or shields, using instrument shelters, adding warning labels, locating stations out of sight from roads, and even employing camouflage. Regular visits to sites allows early detection and repair of any damage so that data loss is minimal.

OPERATING A MONITORING NETWORK

Operating a monitoring system not only involves making regular observations, but also preserving and communicating them. Thus, operation also includes maintaining a database and reporting monitoring results.

Observations

Many types of monitoring observations may now be automatically recorded on data loggers. However, some observations still must be made manually. Using a comprehensive data sheet for recording such observations is important. Data sheets serve at least three purposes. First, they ensure that field personnel consistently make the needed observations by reminding them of the type of data to collect. Second, they provide for continuity, should there be turnover in staff. Finally, they serve as a standard record from which observations may be conveniently entered into a database. Thoughtful design of such data sheets at the outset of a project will save frustration later on. It is easier to ignore superfluous field notes if the data sheet is too detailed than it is to do without observations of basic parameters if it is not detailed enough.

Making careful field observations cannot be over-emphasized. It doesn't matter how good the monitoring network or equipment is, if the field work is sloppy. Become familiar with and consistently use the approved methods for taking measurements and samples. Record everything that is called for on the data sheet. Also, write observations clearly. The lack of, or uncertainty about, the entry for a single parameter may invalidate the entire observation. Developing a routine procedure for tasks when visiting a station may be helpful. As a complement to the standardized data sheet, you should also standardize field work. For example, calibrate instruments, label sample vessels, take measurements etc., at the same time and in the same way on each visit. The evolution of a pattern may minimize omissions or mistakes.

Database

Typically reams of data tables are collected in a monitoring project. All too often these data are never seen by anyone, let alone used in any meaningful way. Ideally, monitoring data should be examined for their environmental significance, reported promptly for immediate use, and archived for future use. Even though a data or technical report has been prepared, results should be filed.

It makes no sense to design and set up a world-class monitoring network and make regular and careful observations if the information gathered is not readily available. Maintaining a database (preferably electronic) is the best way to archive monitoring results and is extremely important. The form of the database depends on what is being monitored. Commercially available software may suffice. It may be necessary, however, to set up a specialized database. Compatibility with the various potential users of the information is the deciding factor. Input from all potential stakeholders or users should be solicited before it is designed. In addition to the client and regulators, this input may also include concerned-citizen groups.

Reporting Results

Observations should not only be entered into a database, they should also be reported. Monitoring results may be communicated informally or formally. An informal report might consist merely of a cover letter, some data tables, and pertinent figures. Such illustrations may include a well-location map, well-construction diagrams, as well as maps or graphs showing water level, the concentration of specific constituents, or the value for selected parameters.

Formal monitoring reports may be required by a government agency. In such a case, the agency may specify the format for reporting. If not, this must be determined by those doing the monitoring. As a minimum, periodic monitoring results should be summarized in a memo to "the file," so they can be recovered later. Whether reporting water-level measurements, thickness of floating product, or results of chemical analyses, it is common to present these observations in tabular form. In such tables it is very useful to give previous values for comparison (see Table 12-3). This information need not include all historical observations (although that is sometimes necessary), but should at least include the values from the previous reporting period and state the change involved.

EVALUATING MONITORING SYSTEMS

Hydrogeologists are often called upon to evaluate monitoring networks. For proposed systems, the monitoring plan is reviewed. For existing systems, the current surveillance operation should be checked against what was proposed or approved in the monitoring plan. Additionally, examine both the current monitoring network and data obtained to date. The ease with which such evaluation may be carried out depends on the information available.

Monitoring Network

A monitoring plan may immediately be rejected if the hydrogeology of the site has not been adequately conceptualized. In such cases additional observation wells and data may be required simply to reach the point where a monitoring plan can be designed, let alone evaluated.

If the site's hydrologic setting has been sufficiently characterized, the evaluation process is very simple. Determine the adequacy of the monitoring network considering the known source(s) of contamination and the conceptual hydrogeologic model. Put another way, are there enough monitoring stations, of the right kind, in the right place, based on what is known of the potential contamination source(s) and the hydrologic system? Thus, monitoring plans should include information on potential sources of contamination (nature and volume) and the hydrogeology of the study area. A single map, showing the potential sources, water-level contours, and the types and location of monitoring stations is essential for evaluating a monitoring plan. If the plan does not contain such a map, it should be requested or constructed, as far as possible, from information available.

Data Set

In addition to evaluating the monitoring network, you must also review the data collected to date. Of particular interest are any trends or anomalies. Although data tables may suggest these, plots of observations (y axis) versus time (x axis) are invaluable. For example, hydrographs may be plotted for stream flow at each station or for water level at each well. Similarly, the concentration of constituents of concern may be graphed over the period of record. Then, the reason for any observed trends or anomalies should be researched and appropriate environmental action taken.

Whereas a single anomalous value may be the result of a sampling or analytical error, multiple anomalous values may be significant. For example, a water-level rise, not readily attributable to natural causes, may indicate a release of liquid waste. Similarly, higher than usual concentrations of a constituent of concern may indicate a release or passage of a plume.

If the change is gradual or characterized by a smooth rise to and fall from the anomalous value, it is probably real. Figure 12-3 shows anomalous tritium values in

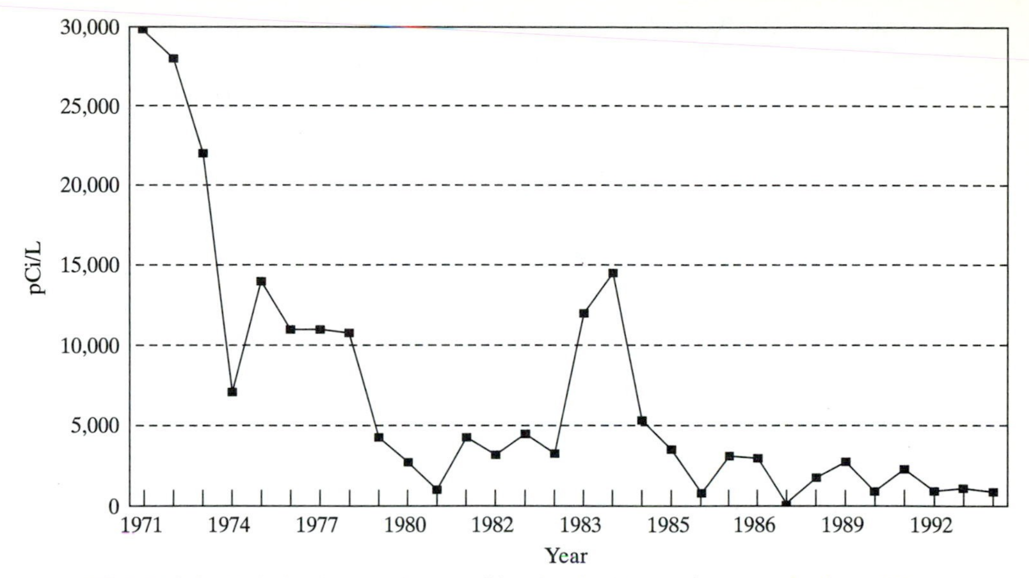

FIGURE 12-3 Variation in ground-water tritium levels with time (measured in picocuries per liter), well LAO 4.5, Los Alamos National Laboratory, northern New Mexico. (Modified from a graph prepared by Michael Dale using Los Alamos National Laboratory Data.)

water from a supply well monitored at Los Alamos National Laboratory in New Mexico. The high value is probably not the result of a lab error, since there is a rise in values preceding it and a drop in values following it on the graph.

If the change is abrupt and persistent, it may indicate a change in sampling method, analytical laboratory used, or some treatment of the well. Figure 12-4 shows an abrupt declining trend in ground-water Trichloroethene (TCE) concentration in a well at Sandia National Laboratories, New Mexico. It turns out that two things happened at the time of the abrupt change. The well was developed to further remove drilling mud, then there was a change in the sampling method. The higher values are associated with samples obtained with a bailer, whereas the lower values are associated with samples taken by a low-flow (Micro-Purge^{tm}) sampler. The implications of such a change are further discussed in the next section.

MONITORING WELLS

The most common component of a surveillance network is monitoring wells, which someone once defined as "holes in the ground that lie." In other words, observations from such wells may not reliably represent what was intended. However, there are ways to ensure that monitoring wells provide more reliable information on site geology, aquifer characteristics, water level, and water quality.

Geology/Aquifer Characteristics

Drilling monitoring wells not only provides observation points for ground-water surveillance, it offers valuable insight into the hydrogeology of the study area. (Similar information is also gained during the drilling of holes for neutron-probe access tubes or

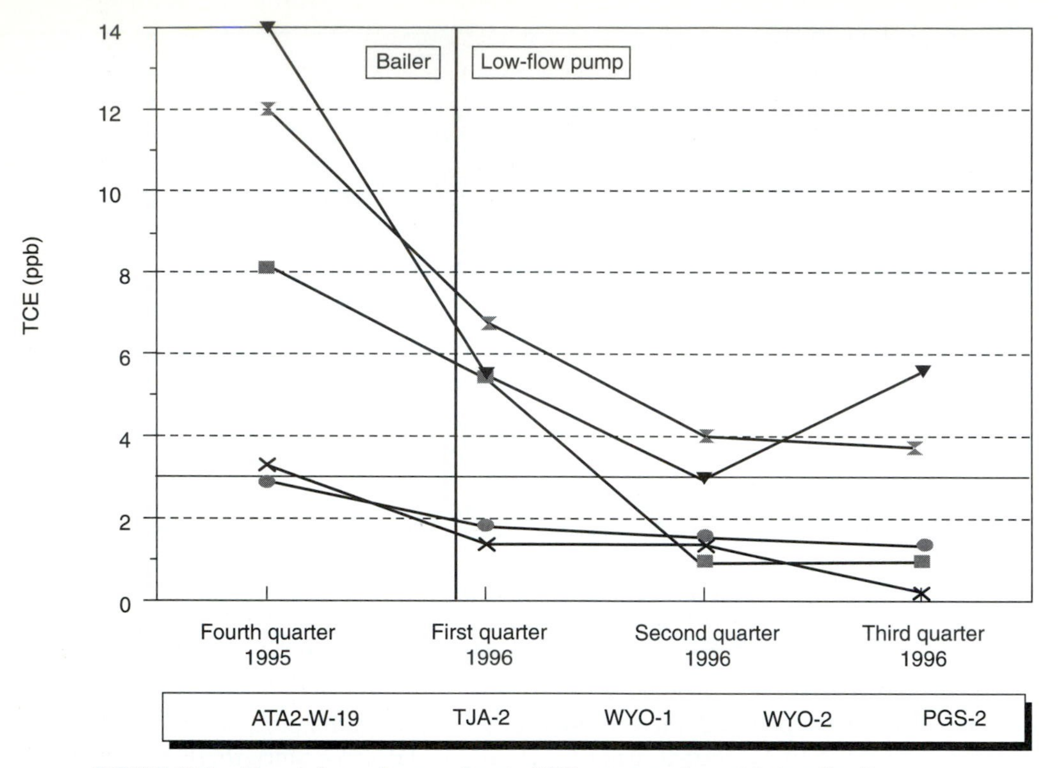

FIGURE 12-4 Abrupt change in ground-water TCE concentrations with time, Sandia National Laboratories, central New Mexico. Note change in sampling method after fourth quarter, 1995. (Modified from a graph prepared by William McDonald using data from Anonymous, 1996.)

samples in soil). Of special interest are the kind of geologic materials present and the position of the water table.

Such information is lost unless a hydrogeologist is constantly on site to oversee the drilling operation. More specifically, the hydrogeologist describes cuttings to characterize and log the materials penetrated, determines the geologic relationship to adjacent wells, and records the depth of saturated intervals encountered. These observations are then used to determine the appropriate well construction and completion.

Water Level

The best way to monitor water level is with piezometers. These wells have very short screen lengths; thus, the water level obtained represents a finite depth in the aquifer. Observations at numerous piezometers located around a site permit the construction of a water-level map. Such maps are commonly used to determine ground-water flow direction and calculate hydraulic gradient.

Monitoring wells have longer screens and are often of larger diameter than piezometers, since they are designed to sample a saturated zone, not merely provide a discrete water level. Since people feel they must have monitoring wells at a site, there are usually plenty of them, but few or no piezometers. Thus, water-level measurements at a site often come from monitoring wells and are averaged over the screened interval. Furthermore, because screens in monitoring wells may be of different lengths and placed at various depths, water-level maps based on measurements in them are only rough approximations. If the direction of ground-water movement is critical, piezometers should be installed so better water-level measurements can be made.

Water-level measurements are also useful in identifying underflow, the water moving in saturated zones beneath stream channels or slopes. Underflow can be an important part of the hydrologic cycle and can be responsible for generating leachate or transporting contaminants (Stone 1997a). Such saturated zones are often perched in unconsolidated material on the bedrock surface, well above the regional water table, which may lie deep within the bedrock. Such saturation may be seasonal, that is, related to higher runoff associated with snowmelt or the rainy season. Piezometers should be installed to document underflow where it is suspected. As such flow is often ephemeral, continuous monitoring and recording is required.

Water Quality

Monitoring wells are used mainly to obtain samples for characterizing ground-water chemistry at a site. Considering the impacts that the installation and monitoring of wells can have on aquifers, obtaining a representative sample from them may be an impossibility (Pennino 1988). This "well trauma," as it may be called, includes disturbances caused during drilling, construction, development, purging, and sampling. Anomalous analytical results can of course be due to laboratory error. More important, however, may be impacts of well construction and the sampling method.

Obviously, we must sample ground water and monitoring wells are the only means we have (Figure 12-5). Many problems can be eliminated by proper well construction. The EPA manual on monitoring wells (Aller and others 1991) provides detailed guidance on well construction. Although it is beyond the scope of this chapter to offer the complete discussion given there, the procedures may be generalized in just a few steps:

1. Characterize the geologic material penetrated by examining cuttings, core, and geophysical logs.
2. Characterize the hydrologic conductivity of the material penetrated based on lab tests, textbook values, pumping tests, or modeling.
3. Keep the screen short (long screens promote vertical movement and mixing of waters in the annulus, perhaps of different qualities, with the resulting sample not being representative of any specific horizon).
4. Use a screen-slot and filter-pack size appropriate for the grain diameter of the aquifer (see Aller and others 1991).
5. Isolate the screened interval from possible contamination from the surface with an adequate annular seal.

Further information on drilling and constructing wells is given in Chapter 10.

In spite of proper well construction, anomalies can be introduced by the sampling method itself. For example, purging (removing stagnant bore water) or even the inser-

FIGURE 12-5 Typical monitoring/recovery well at a pipeline leak site, Endee, New Mexico.

tion of a sampler causes turbulence and thus turbidity. It has been shown that the concentration of metals is elevated in samples taken with non-dedicated samplers (Puls and Paul 1995). This elevation is due to the increase in particulate matter to which metals tend to adhere.

Appropriate matching of the diameter of filter-pack material and the screen-slot size to the grain size of the formation, together with adequate well development, should essentially eliminate the influx of particulates. However, sometimes it is inevitable. Thus, alternative sampling methods, that minimize turbidity, are always being developed and tested.

The use of dedicated, low-flow samplers has received a lot of attention because they not only reduce the turbulence that leads to turbidity, but also the volume of purge water that must be disposed of as well. A concern with "micropurging," as it is sometimes called (after the trade name of one low-flow sampling system, Micro-Purge^{tm}), is its assumption that the water sampled comes from the formation directly opposite the intake. Although various studies involving tracers and tracking suspended particles by means of the colloidal borescope have been conducted to test this assumption, the source of samples obtained by low-flow samplers is still not entirely clear (Stone 1997b). The recommended protocol is to set the low-flow pump just above center of the screened interval. However, as hydraulic conductivity varies within even a seemingly homogeneous hydrogeologic medium, so, too, will its ability to contribute contaminant to the well.

Consider the case mentioned above, where the concentration of TCE dropped when the method of sampling was changed from bailing to the low-flow method. Paradoxically, the low-flow sample from a discrete portion of the screened interval had a lower concentration of TCE than did a bailer sample, which is a blend of water from all of the screened interval. This drop in TCE content was attributed to removal of drilling mud by further well development. However, the discrepancy may be better explained by the fact that the low-flow sampler was placed opposite a low-permeability horizon.

Thus, two things should be done before deciding where to set the sampler.

1. If possible, determine the vertical distribution of hydraulic conductivity for the screened interval (this may be estimated from the lithologic characteristics indicated by geologic or geophysical logs of the well).
2. Operate the sampler at regular intervals over the entire screen length or in each zone of contrasting productivity (hydraulic conductivity); and plot and examine a vertical profile of concentrations for the constituent of concern.

Then place the pump to sample the zone yielding the maximum contaminant concentration.

CASE HISTORIES

Examining some existing monitoring programs is a good way to illustrate their basic features. Such systems vary in scale and complexity, as well as in monitoring target and procedures. These examples were selected to demonstrate this variation. While the first case is fairly simple, and the site is of medium size, the second case is complex and involves a very large site.

AT&SF Railroad Yard, Belen, New Mexico

Spillage over a period of time at this diesel-locomotive fueling facility and switching yard resulted in the accumulation of floating product on the shallow water table. This was first detected in a drainage ditch, to which the people in the neighborhood remember delighting in setting fire as kids.

Site Characterization

The site lies in the Albuquerque-Belen Basin of the Rio Grande Rift and is underlain by basin-fill sediments. The interval between the surface and the water table is characterized by sand, silty sand, and clay. The water-table is very shallow, with depth generally averaging <10 ft. The aquifer is unconfined and consists of gravel, sand, silt, and clay, with texture varying markedly over short lateral distances. Results of a pumping test suggest hydraulic conductivity is on the order of 10^{-2} cm/sec, which is representative of clean or silty sand. Ground-water flow is to the south, with an average hydraulic gradient of approximately 5.5 ft/mi.

As a result of site studies, three plumes of floating product are recognized. The northern plume is the smallest, being approximately 600 ft long and 400 ft wide, with a maximum thickness of approximately 2 ft. The central plume is the largest, with a length of approximately 1,900 ft and a width of 150 – 600 ft (Figure 12-6). Maximum thickness of the central plume is approximately 6 ft. The size of the southern plume is intermediate, having a length of approximately 730 ft, a width of approximately 350 ft, and a thickness of approximately 6 ft. A comparison of some historical data for floating product thicknesses in some of the monitoring wells is given in Table 12-3.

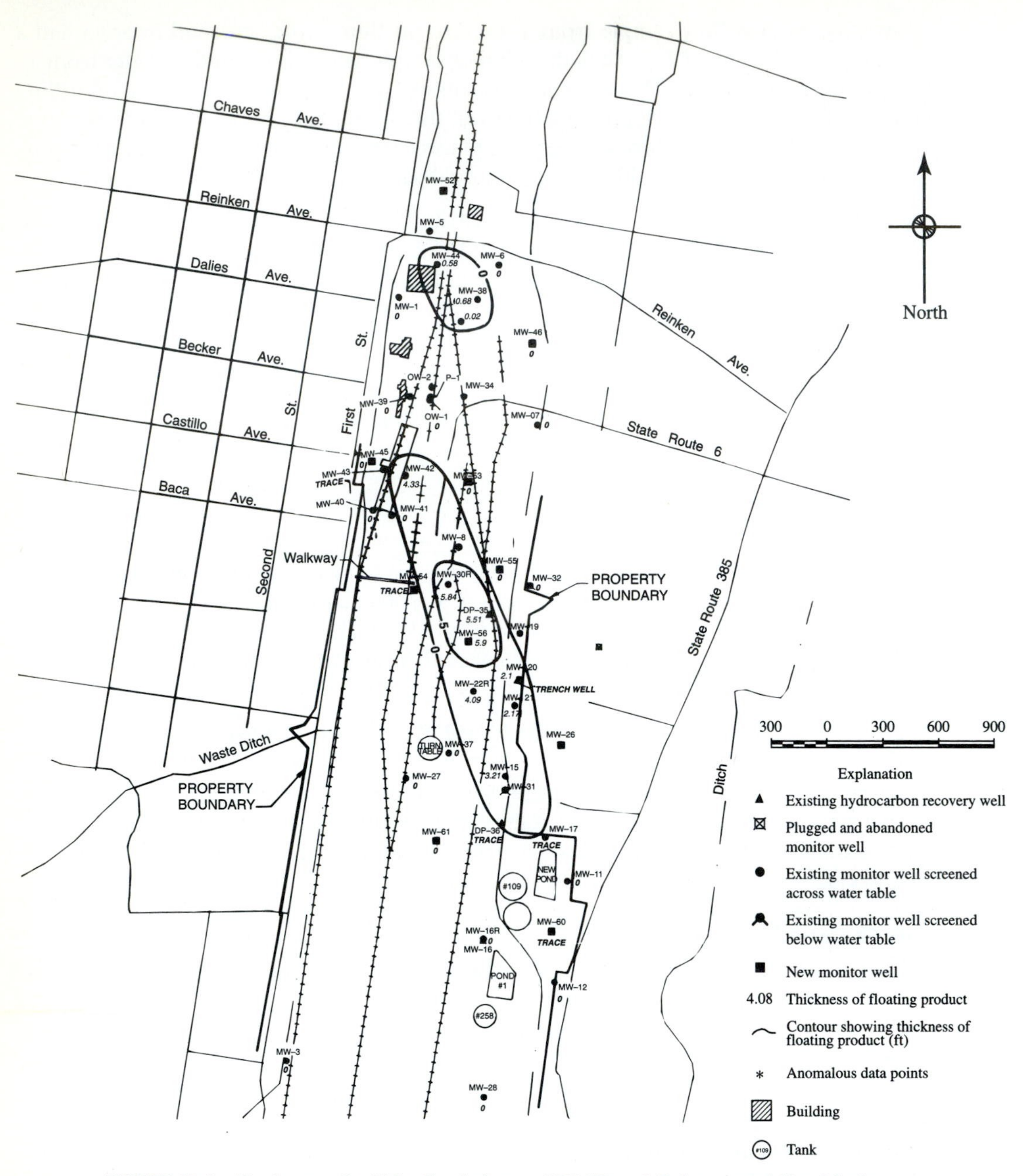

FIGURE 12-6 Northern and middle diesel plumes, AT&SF yard, Belen, central New Mexico. (Modified from Radian Corporation 1992, Figure 3-2.)

Monitoring Target, Focus and Objective

The target here is ground-water. The focus is twofold: ground-water movement and quality. More specifically, fluid levels are regularly measured to monitor the thickness of floating product and ground-water flow direction. Additionally, ground water is sampled regularly to monitor dissolved petroleum hydrocarbons. The objective is to characterize the extent and movement of the plumes in order to most efficiently remediate them.

Monitoring Network

The monitoring network for the entire property (all three plumes) consists of 79 wells. Fluid levels (water and product) are measured manually in all of them on a quarterly basis. A smaller subset of wells is used for water-quality monitoring. More specifically, samples from 10 wells are analyzed biquarterly (first and third quarters) for BTEX (benzene, toluene, ethylbenzene, and xylene), chlorobenzene, dichlorobenzene, and PAH (purgeable aliphatic hydrocarbons).

Discussion

Extensive remediation is under way at the central plume, where an innovative approach to remediation is employed due to site conditions. Because of the low-permeability of the contaminated sediments, free product is removed by means of recovery wells constructed in a collection trench filled with more permeable material (see Figure 11-3). Ideally, such a trench would be constructed east-west, perpendicular to the southerly ground-water flow direction. However, because the tracks are aligned north-south, an east-west trench was not possible. Thus, a north-south trench was positioned just inside and parallel to the eastern railroad property boundary. There it prevents the plume from migrating off-site. Approximately 189,000 gal of hydrocarbons had been recovered as of September 1995.

It is also interesting to note that the monitoring process here is quite dynamic. That is, if results indicate a change in product thickness or ground-water quality, new wells are often installed to determine whether the change is due to movement of the existing plumes or if there is a new source. Although this is a good philosophy for any site, it does not always prevail for various reasons. Also, existing monitoring wells are frequently rendered useless by the day-to-day operations of the railroad and must be replaced. Some are damaged when vehicles or construction equipment hit them, some are buried by fresh ballast material, and others are overrun by new tracks! This apparently happens because the daily-operations and environmental sides of the company are located miles apart, have different missions, and don't communicate enough. Unfortunately, that is not unique to this case and can cause problems anywhere.

Carlin Gold Trend, Nevada

Monitoring is important at open-pit gold mining operations, such as those addressed here, because they have the potential for impacting both water quantity and quality. Pumping ground water for dewatering the pits lowers water levels. Leakage from tailings impoundments and heap-leaching operations can contaminate ground water, if it goes undetected.

Site Characterization

The Carlin Trend is located in typical Basin-and-Range country of the American West and lies in or adjacent to three hydrographic basins. These basins are filled with Tertiary/Quaternary alluvial and lacustrine sediments and volcanics, whereas the ranges are geologically complex and consist of Paleozoic limestone and siltstone, both of which are fractured and faulted. Ground water in the basins is generally unconfined and flows southerly toward the Humboldt River, the main perennial stream in the region.

Monitoring Target, Focus and Objective

The monitoring plan described here is that originally used in the area (Stone, 1991). As no formal monitoring was conducted prior to implementation of this plan, no plumes, trends or anomalies had been identified.

The monitoring was designed to target both surface water and ground water. For surface water, the focus was detecting changes in discharge and water quality. For ground water, the focus was detecting changes in water level and water quality. The objective of monitoring, in both cases, was to identify changes in parameters as an aid to early detection and remediation of any hydrologic impact of mining.

Monitoring Network

The monitoring network included 95 stations devoted variously to surface- and ground-water surveillance (Table 12-4). Surface-water quantity (discharge) was monitored at 9 stations on six streams. Discharge was monitored continuously by means of six recording stream gages (two on

TABLE 12-4 Distribution of Stations by Monitoring Function, Maggie Creek Basin, Carlin Trend, Nevada[a]

	Surface-Water Monitoring Stations		
	Discharge		
Stream	Recorder	Flume	Water Quality
Jack Creek	1	0	1
Simon Creek	0	1	1
Maggie Creek	2	0	4
James Creek	0	1	0
Marys Creek	1	0	
Humboldt River	2	0	8
Total Stations	6	2	14

	Ground-Water Monitoring Stations	
Aquifer	Water Level	Water Quality
Alluvium	3	3
Volcanics	3	0
Basin Fill	16	6
Siltstone	11	6
Limestone	9	2
Total Stations	42	17

[a]Analytes for water quality were as suggested by the state, and were different for surface and ground water.

Source: From Stone 1991.

FIGURE 12-7 Flume for monitoring discharge of an ephemeral stream, Carlin Gold Trend, northern Nevada.

the perennial stream) and monthly by means of two flumes (Figure 12-7). Surface-water quality was monitored on a quarterly basis by analyzing samples from 13 stations, 9 of which were on the perennial stream. Ground-water level was monitored quarterly at 27 wells (3 in the volcanics, 11 in the basin fill, 5 in the siltstone, and 8 in the limestone). Ground-water discharge was measured semiannually (spring and fall) at 43 springs. Ground-water quality was monitored monthly at 5 wells (2 completed in the basin fill, 2 in the siltstone and 1 in the limestone) and semiannually at 9 springs. Well samples were taken by small-diameter, dedicated submersible pumps (Figure 12-8).

FIGURE 12-8 Small diameter submersible pump dedicated to water-quality monitoring wells, Carlin Gold Trend, northern Nevada.

Discussion

Monitoring here is unique because surveillance duties were initially divided among three parties: the two mining companies working in the area and the U.S. Geological Survey, whom the state engineer designated to provide oversight, under funding by the mining companies. Although each of the parties had a specific area for which they had monitoring responsibilities the areas overlapped, so that each party could check their results against those of at least one other party.

This site also illustrates that monitoring is a dynamic process. As is often the case, it was proposed that the number of stations in the network and the frequency of monitoring be reduced after the plan had been operated for a year or so. As mining progresses, some wells are inevitably eliminated and others installed.

Note that adapting monitoring methods to local conditions was cost effective. This was especially true for surface-water measurements. The low-technology equipment for monitoring streamflow in this arid setting was considerably cheaper than continuous recording gages.

REFERENCES

Aller, L., T. W. Bennett, G. Hackett, R. J. Petty, J. H. Lehr, H. Sedoris, and J. E. Denne. 1991. *Handbook of suggested practices for the design and installation of ground-water monitoring wells.* Report EPA/600/4-89/034, U.S. Environmental Protection Agency, 221 p.

Buchanan, T. J. and W. P. Somers. 1968. *Stage measurement at gaging stations.* In *Techniques of Water-Resource Investigations of the United States Geological Survey,* Book 3. Chapter A7, U.S. Geological Survey. 28 p.

Carter, R. W. and J. Davidian. 1968. General procedures for gaging streams. In *Techniques of Water-Resource Investigations of the United States Geological Survey,* Book 3, Chapter A6. U.S. Geological Survey. 13 p.

Everett, L. G., L. G. Wilson, and E. W. Hoylman. 1984. Vadose zone monitoring for hazardous waste sites. *Pollution Technology Review* 112. Noyes Data Corporation, Park Ridge. NJ. 360 p.

Heath, R. C. 1976. Design of ground-water level observation-well programs. *Ground Water* 14(2):71–77.

Kazmann, R. G. 1981. An introduction to ground-water monitoring. *Ground Water Monitoring Review* 1(1):28–30.

Nielsen, D. M., and A. I. Johnson, eds. 1990. Ground water and vadose zone monitoring. STP 1053. ASTM, Philadelphia, PA. 313 p.

Pauszek, F. H. 1972. Digest of the catalog of information on water data. Open-file report. U.S. Geological Survey. 66 p.

Pennino, J. D. 1988. There's no such thing as a representative ground-water sample. *Ground Water Monitoring and Remediation* 8 (3): 4–9.

Puls, R. W. and C. J. Paul, 1995. Low-flow purging and sampling of ground water monitoring wells with dedicated systems. *Ground Water Monitoring and Remediation* 15(1): 116–23.

Radian Corporation. 1995. Remedial monitoring report, third quarter 1995, AT&SF fueling facility and switching yard, Belen, New Mexico, final. Consulting report. Radian Corporation, Albuquerque, NM.

Stone, W. J. 1991. Maggie Creek Basin monitoring plan. Hydrology Department Report. Newmont Gold Company, Carlin, Nevada.

Stone, W. J. 1997a. Importance of underflow in environmental site characterization (abs). *Proceedings.* Annual Spring Meeting, New Mexico Geological Society.

Stone, W. J. 1997b. Low-flow ground-water sampling—is it a cure-all? *Ground Water Monitoring and Remediation.* 17(2): 7–72.

Summers, W. K. 1985. Conceptualization of ground-water flow systems and the design of monitoring programs. Preprint no. 85-365. Society of Mining Engineers of AIME. 10 p.

U. S. Environmental Protection Agency. 1986. RCRA ground-water monitoring technical enforcement guidance document (TEGD). Report OSWER-9950.1. U.S. Environmental Protection Agency. 208 p.

U. S. Geological Survey. 1977. *National handbook of recommended methods for water-data acquisition.* Reston, Virginia: U.S. Geological Survey.

CHAPTER 13

Water Problems In Mining

The water problems associated with mining are many and varied. Both surface and underground mines require dry working conditions, and dewatering may be required. Because water is used in many ways in mining, an adequate supply must be obtained. The chemical quality of any dewatering product, as well as that of the potential supply water, may be problematic. It is also possible that mining operations may impact the quantity and quality of the regional water resources.

Because the cost of resolving such water problems can be an important factor in determining whether or not mining a mineral deposit is economically feasible, various types of hydrogeologic studies are commonly conducted along with developing mining operations. Many water problems can be anticipated, or even avoided, if a regional hydrogeologic study is conducted during the initial assessment of a potential mine property. For example, preliminary studies proved useful at the Fence Lake coal mine in west central New Mexico. Because federal surface-mining regulations call for restoring strip mines to their original recharge capacity, Salt River Project, the leaseholder, supported a pre-mining study of ground-water recharge rates (Stone 1984). In response to dewatering and supply concerns, an investigation was also conducted to formulate a conceptual hydrogeologic model of the area (McGurk and Stone 1986).

In a preliminary study, the level of investigation depends on the level of interest in the site, but at least basic information should be compiled. Such a study can then be followed by more specific studies as needed. However, it is necessary that as much should be learned about the hydrogeologic system as early as possible, if mining is to proceed.

Some types of hydrogeologic information may also be required by various state regulatory agencies before mining commences. Thus, it is a good idea to check with them before getting too far into the hydrogeologic study planning process. In some states a publication may be available that summarizes the requirements for developing a mine. This is especially useful where different agencies regulate different aspects of mine operations. For example, in New Mexico, a single booklet has been prepared that gives the various regulatory standards, including those involving water issues, that a company must meet before a mine permit can be issued (New Mexico Energy, Minerals, and Natural Resources Department 1990). If mining is a significant industry in your state, but such a publication does not exist, you may want to suggest one or see to it that one is prepared.

Water problems in mining receive considerable attention worldwide. In some cases separate professional organizations, such as the International Mine Water Association, organized in Spain, have been formed to address them. Special publications on water problems in mining prepared by these and other professional organizations are an excellent source of information. Examples include the proceedings of a symposium on water problems in mining sponsored by the American Water Resources Association (Hadley and Snow 1974), selected papers on mine-water control by the Hungarian Mining Development Institute (Kesseru 1979), a manual on mine abandonment for operators in Spain by the International Mine Water Association (Fernandez-Rubio and others 1986), and a series of modules on best-practice environmental management in mining produced by the Australian Environmental Protection Agency (Australian Federal Environment Department 1995).

DEWATERING

If the zone to be mined is found to be saturated, dewatering will be required. Dewatering projects consist of three main components: characterization, system design, and disposal-option review. Although the system design, including installation and maintenance of the dewatering system, is mainly an engineering problem, the other two components require hydrogeologic studies.

Characterization

Before a dewatering system can be designed, the water-producing zone(s) must be fully characterized. This characterization applies to both the hydrologic system and its response to pumping. More specifically, the aquifers, their thickness and extent, their hydraulic properties, water-level depths, and elevations, as well as drawdown for various production rates, must be determined. Because selection of a disposal method will require information on the physical and chemical quality of the dewatering product, this should be compiled during the characterization phase as well. For example, is the temperature of the water elevated? Does it contain toxic constituents?

The ease with which characterization is accomplished depends on the availability of previous hydrogeologic studies. If an initial regional water study was made, a conceptual hydrogeologic model may already exist. This presumably includes the water-level maps, hydrogeologic cross sections, water-quality maps, as well as supporting well logs, well-records, and water-quality data needed to design a dewatering system. If an initial hydrogeologic study was not conducted, the basic information must be obtained elsewhere. This process may simply involve compiling geologic and hydrologic information from various sources, as described in earlier chapters. If no previous studies exist, field work will be required. This may include drilling test wells, performing pumping tests, and collecting ground-water samples for analysis. Finally, the information gained from these activities must be synthesized into a conceptual hydrogeologic model, especially as it applies to the dewatering.

Numerical models are an important tool in the characterization phase of dewatering studies. Models are often used to characterize the steady-state ground-water system. Once this is done, the effectiveness of various dewatering schemes can be evaluated with transient models. Since the method of disposing of the dewatering product may dictate some aspects of the pumping system, modeling may include various disposal scenarios as well. Models may also be used to predict the impact of dewatering on the regional hydrologic system. (Ground-water modeling is the subject of chapter 14.)

Product Disposal

Depending on site conditions, a large volume of water may be produced in dewatering. This must be disposed of in an environmentally sound and economically feasible manner. One or more of various options may be feasible:

reinjection
mine water supply
infiltration
wetlands
irrigation of pasture or crops
treatment and release to a local river or lake

There are advantages and disadvantages to each disposal method. An option's ultimate suitability depends on site-specific conditions. It is wise to learn the state's general position on the various options before investing much time or money in feasibility studies. Selection of the best disposal method usually requires additional hydrogeologic study.

Regardless of the disposal method considered, the volume and water quality of the dewatering product must be known. If data from regional investigations are inadequate, site-specific studies may be required. Similarly, all disposal options will require a means of storing some volume of dewatering product for some period of time when it is necessary to interrupt disposal operations. If the use of tanks is not possible, the hydrogeologist must make geotechnical studies to determine potential sites for surface impoundments or reservoirs.

Additionally, some disposal methods may require special information. For example, water levels and hydraulic conductivities are required for potential reinjection intervals. Obviously, the best target for reinjection is a highly permeable zone. However, that is often the zone that is being dewatered, and reinjection would be counterproductive. For infiltration, the unsaturated volume of target materials must be known. Potential evapotranspiration rates and uptake capacity of plants must be known for the wetlands option (Figure 13-1). Agricultural disposal methods require the irrigation potential of area soils and water application rates for suitable crops.

FIGURE 13-1 Wetlands constructed to remove metals in dewatering product at the Sleeper Gold Mine, northern Nevada.

The hydrochemistry of dewatering product may limit the disposal options that may be possible without water treatment. For example, water pumped to keep a mine dry may contain elevated levels of objectionable constituents. Usually, these may be readily removed by chemical precipitation. However, in some cases, they are in a form that is more difficult to treat.

Hydrochemistry and temperature of both the dewatering product and receiving waters, as well as surface-water–quality standards, must be determined if water is to be treated and released. Bench and field tests to determine treatability may also be required if the hydrochemistry of the dewatering product is unusual.

WATER SUPPLY

Water is used for many purposes around a mine. Of course, there must be a potable water supply for employees, both at work and in any associated company town, if it is located at a remote location. A supply of water is also needed for such diverse uses as drilling, dust control, washing equipment, milling, smelting, boiler-feed and cooling (in the case of coal-fired, mine-mouth power plants), irrigation during reclamation, and fire-fighting. This water need not be potable.

As noted in the previous chapter, the two main concerns in water supply are quantity and quality. Thus, investigations are required to determine if local sources can reliably produce both the quantity and quality of water needed.

Quantity

Quantity issues are twofold: how much water is needed and how much is available? Is the local aquifer productive enough to support a single high-volume well or will it take a network of widely spaced wells to produce what is needed? What is the anticipated life of the local supply for the production levels anticipated? Ground-water modeling is the best tool for addressing such questions. Once a steady-state model is developed, it can be stressed at the projected pumping rate to determine drawdown and its impact on any nearby private wells.

Quality

Quality requirements vary depending on how the water will be used. Potable supplies must meet drinking-water standards. An abundant supply may be provided by dewatering, but it may not be potable. However, water to be used for various mine applications need not meet drinking-water standards. Nonetheless, the quality of nonpotable water can also be important. Water for use in milling, boiler feed, etc. need not be of drinking-water quality, but elevated concentrations or even the presence of certain constituents may be objectionable. Since water treatment is an additional cost, the extent to which it may be required for various uses must be determined.

Where there is no dewatering or where the product is of inadequate quantity or unsuitable quality, the location of a suitable water supply may require additional hydrogeologic study. If such a study was made during the initial evaluation of the property or for dewatering, much of the information needed may be already available. If not, it must be compiled from outside reports (as described in chapters 2 and 5) or obtained through drilling, pumping tests, and water analysis.

If locally available ground water cannot be used for a potable water supply, an alternative must be located. This can be very expensive. For example, although the quality of water in the discharge area of the Great Artesian Basin is not good, it is better than that of water at the Roxby Downs mine in South Australia. Thus, water used at the mine is piped from wells many kilometers away (Figure 13-2).

IMPACT OF MINING

Another important issue is the impact mining operations can have on the regional water resources. Such impacts fall into the two basic categories previously considered: water quantity and water quality. Special hydrogeologic studies may also be required to address these impacts.

Water-Quantity Impacts

Mining may affect the quantity of regional water resources in various ways. The major impacts are to the local or regional water balance or to water levels. Land-use changes associated with mining can alter the water balance by changing various components of the hydrologic cycle. For example, where surface mining disrupts streams, runoff is reduced. When dewatering product is discharged to area streams, runoff is increased. The

FIGURE 13-2 Water-supply well for the town site, Roxby Downs Mine, South Australia.

destruction of vegetation in surface mining reduces evapotranspiration and increases recharge. The creation or destruction of depressions during regrading for reclamation can increase or decrease recharge.

Dewatering in both surface and underground mines, together with pumping of supply wells, invariably lowers the water table or potentiometric surface. This may have various consequences. The most obvious impact is a water-level drop in or even drying up of public or private wells in the immediate vicinity of the mine. Similarly, area spring flow may be reduced or cutoff entirely. Because hydraulic head controls ground-water movement, a more serious impact is a change in the amount or direction of regional ground-water flow. Since ground-water systems often discharge to major rivers, water-level declines associated with dewatering can also impact stream flow. Because such discharge areas can be far from the mines, dewatering can have far-reaching consequences.

Water-Quality Impacts

Mining may impact the quality of regional water resources as well. This may involve contamination by various by-products of the extractive process.

Acid mine drainage is a common water-quality problem in many mining areas. This drainage consists of the low-pH waters generated in areas of metal and coal mining. Sulfuric acid is produced by the reaction of sulfide minerals in the ore with oxygen and water. The acid produced by the oxidation of pyrite can, in turn, dissolve other metals that are undesirable in runoff.

Even abandoned mines can produce large volumes of acid runoff. Some such runoff may be associated with waste-rock piles containing sulfide minerals. This is the case at the Terrero Mine, an abandoned underground lead-zinc operation, along the east bank of the Pecos River, north of Pecos, New Mexico (Figure 13-3). Seeps at the base of the waste-rock pile are characterized by low pH and elevated concentrations of lead, zinc, and copper. Such seepage has killed vegetation and may have affected fish in the river, including those at a hatchery 11 mi downstream (Adrian Brown Consultants 1991).

Especially problematic are tunnels constructed to drain metals mines, because they continue to function long after mining has ceased. For example, the Argo Tunnel in Idaho Springs, Colorado, (Figure 13-4) was completed in 1904 and continues to discharge acid to Clear Creek, although mining operations have ceased (Boyles and others 1974). Aquatic life is not only impacted by the low pH, but also by harmful concentrations of heavy metals. Similar problems are associated with the extraction of high-sulfur (high-pyrite) coal, regardless of mining method.

Studies are also required to determine the quality of any standing water that will accumulate in pits once dewatering ceases and the property is abandoned. For example, the chemistry of the water in the lake that occupies the Berkeley Pit at Butte, Montana, following cessation of dewatering in 1982 has been investigated in some detail (Camp, Dresser, and McKee 1988). The objectives of the study were to characterize the distribution and concentration of metals and to model the feasibility of neutralizing pit water using an alkaline tailings solution produced on-site. More specifically, the study evaluated five scenarios, including two in which the lake surface was frozen over, to represent winter conditions. Results showed that addition of alkaline slurry would alter pH so that major metal species would be reduced by precipitation, sorption, and chelation. For the volume investigated, treatment would take 10 years.

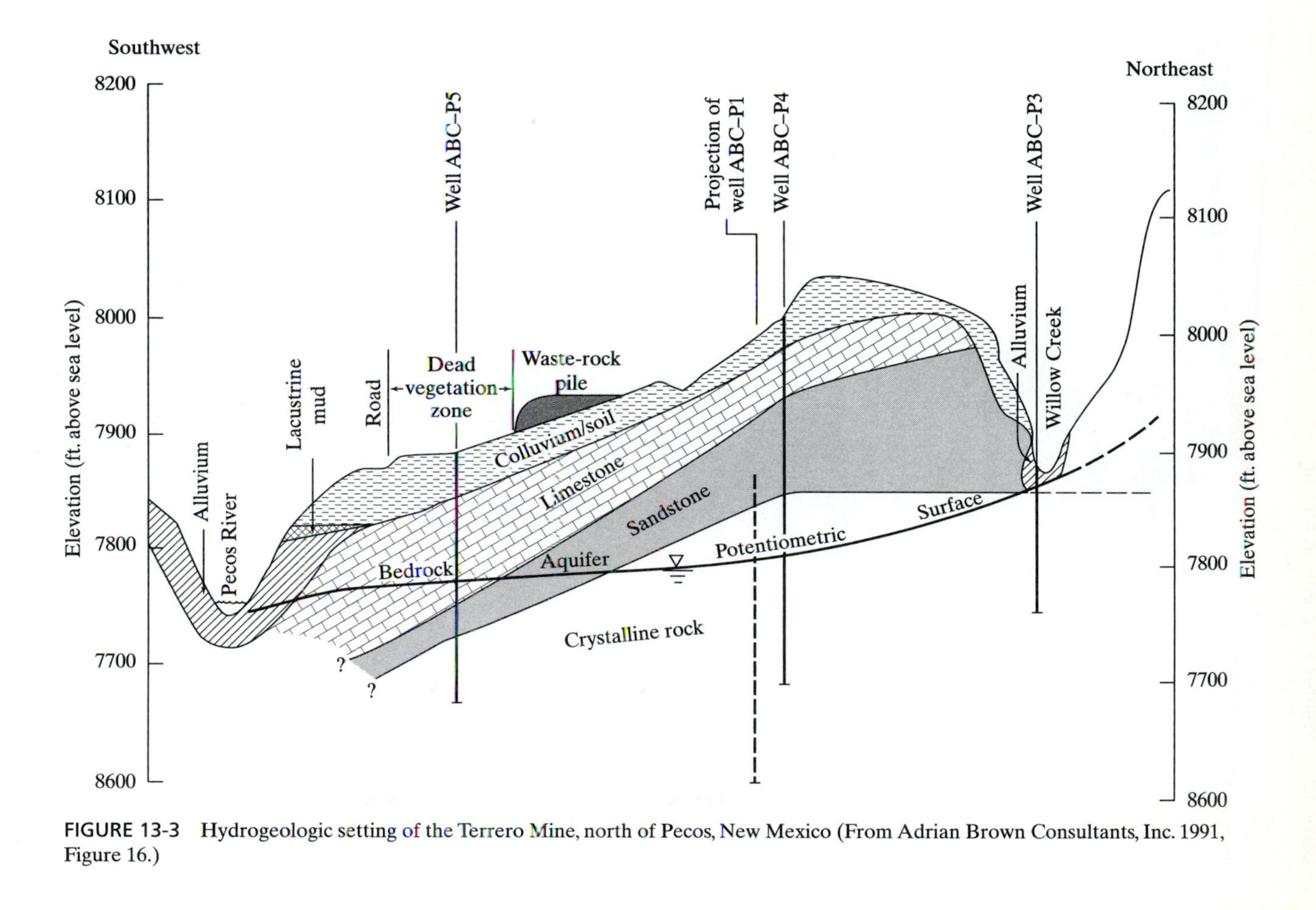

FIGURE 13-3 Hydrogeologic setting of the Terrero Mine, north of Pecos, New Mexico (From Adrian Brown Consultants, Inc. 1991, Figure 16.)

FIGURE 13-4 Acid mine drainage from Argo Tunnel, Idaho Falls, Colorado.

Another subject for hydrogeologic study is any degradation in the quality of water pumped from public or private wells near mines. The most obvious cause of such change is the release of waste or contaminants at the mine. Investigation should focus on possible leaks or spills around chemical and fuel storage tanks, pipelines, and loading ramps, as well as drainage and tailings ponds. A less obvious cause of water-quality impact is mixing of ground waters of good and poor quality due to a change in head associated with dewatering. This is most notable where there are multiple confined aquifers of differing water quality.

CASE HISTORIES

The following examples of hydrogeologic studies conducted at mines or in mining areas are offered to show the kinds of water problems encountered and how they are addressed. Since the summaries are necessarily brief, readers are directed to the original references, if more detailed information is desired.

Grants Uranium Belt, New Mexico

The Grants Uranium Belt parallels the outcrop of the Morrison Formation (Jurassic) along the southern margin of the San Juan Basin, a large structural depression in northwestern New Mexico. Prior to a drop in the price of "yellow cake" and a general moratorium on nuclear power plants in the 1980s, uranium was extensively mined from the Morrison by underground methods throughout the Grants Uranium Belt. Because the Morrison is also a major artesian aquifer in the region, dewatering was required.

In response to concern over the regional impact of dewatering associated with this mining, the state engineer called for a complete hydrogeologic study of the basin. The purpose of the study was twofold: (1) to provide a conceptual hydrogeologic model of the basin for the state engineer's use in evaluating water-rights applications for dewatering, and (2) to evaluate the potential long-term impact of uranium-mine dewatering on ground-water discharge to the San Juan River and Rio Grande. The study was set up as a 5-year project to be cooperatively undertaken by the USGS and the state geological survey (New Mexico Bureau of Mines and Mineral Resources).

Conceptual Hydrogeologic Model

The basin contains more than 14,000 ft of sedimentary rocks, ranging in age from Paleozoic through Cenozoic (Stone and others 1983). Since impact of uranium mining in the Morrison was the focus, the study centered on sedimentary strata of Jurassic and younger age. This sequence includes numerous confined sandstone aquifers separated by shale aquitards. Geologic cross sections were made using the geophysical logs afforded by numerous petroleum wells in the basin. Such logs and well data also supported the preparation of structure, depth, and thickness maps for each sandstone. Hydrologic data permitting, maps of water level, transmissivity, and water quality (specific conductance) were also prepared for each aquifer. From these maps regional ground water was conceptualized as flowing from recharge areas in the uplifts surrounding the basin toward three main discharge areas: the San Juan River in the north, tributaries of the Little Colorado River in the southwest, and tributaries of the Rio Grande in the southeast.

Local Impact of Dewatering

The quality of water in some private wells in the uranium-mining area was noted to have deteriorated over time. A good piece of hydrogeologic detective work by Kelly and others (1980) determined the cause. Pre-mining conditions were characterized by an upward vertical gradient in the artesian sandstone aquifers: Head was higher in the Morrison Formation than in the overlying Dakota Sandstone. Because water in the Dakota is of poor quality and that in the Morrison is of good quality, most area supply wells tapped the Morrison. However, dewatering associated with mining lowered the potentiometric surface for the Morrison such that water from the overlying Dakota Sandstone flowed down into it (Figure 13-5). Eventually, Morrison wells produced undesirable Dakota water (Figure 13-6).

Regional Impact of Dewatering

As uranium-mine dewatering produced large quantities of water from a regional aquifer, a special study was conducted to address the extent of potential water-level declines and reduction of ground-water flow to the San Juan River and the Rio Grande (Lyford and others 1980). This work

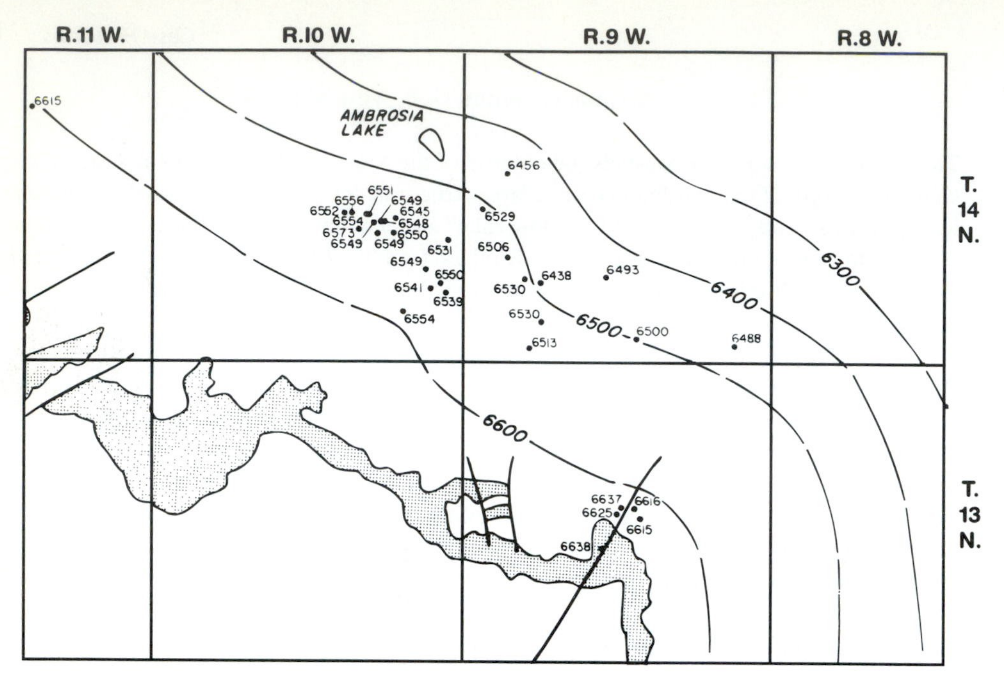

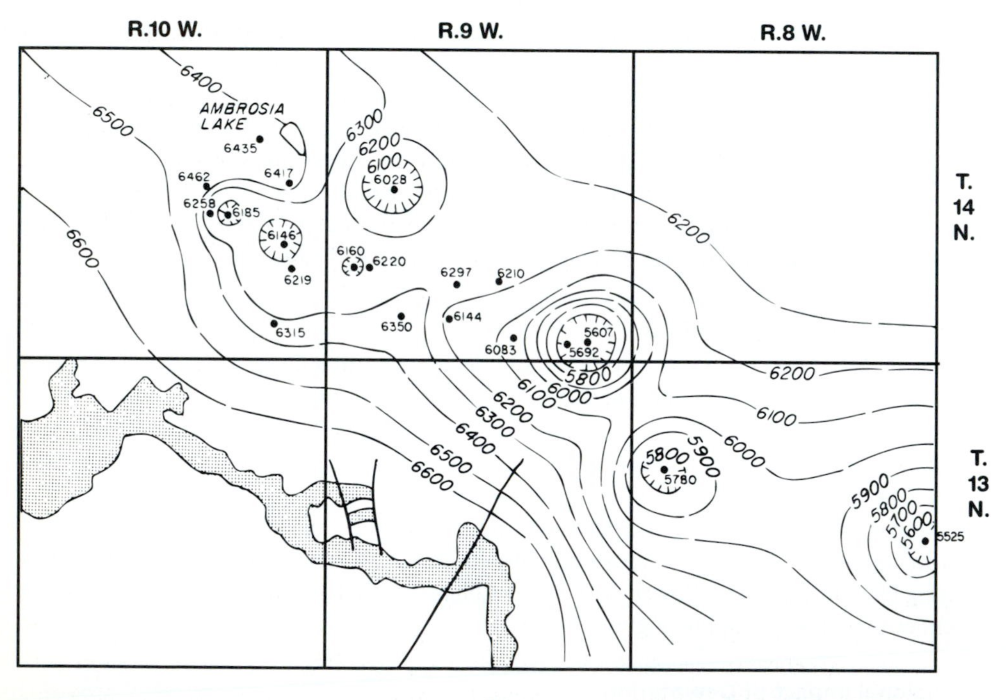

FIGURE 13-5 Potentiometric surface for the uranium-bearing sandstone (Westwater Canyon Member of the Morrison Formation) in the Ambrosia Lake area, near Grants, New Mexico; (a) pre-mining condition, and (b) extensive-mining and dewatering condition. Stippled pattern is the outcrop of the Morrison Formation. (From Kelly and others 1980, Figures 4 and 6.)

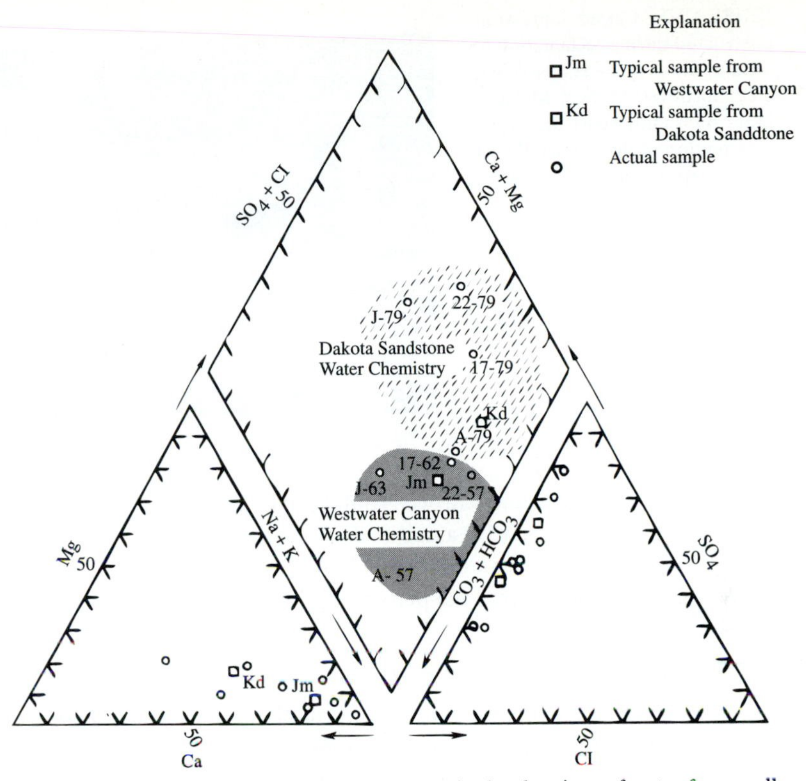

FIGURE 13-6 Piper diagram showing a change in the chemistry of water from wells completed in the Westwater Canyon Member of the Morrison Formation due to mine dewatering. (Note: The year of the data is indicated by the numerical suffix in the label. J-79 indicates data from 1979, for instance.) Early analyses (in the 1950s and 1960s) plot in the water-chemistry field for that aquifer, but later analyses (1979) plot in the field for the overlying Dakota Sandstone. Pumping reversed the gradient, allowing Dakota water to enter Westwater wells. (From Kelly and others 1980, Figure 9).

is a good example of the kind of study that may be required in mining areas. (However, its presentation is reserved for Chapter 14, where it is a case history for modeling.)

Navajo Coal Mine, New Mexico

The Navajo Mine also lies in the San Juan Basin, but in the northeastern part dominated by Cretaceous coal measures. The section is characterized by coal beds, alternating with marine and nonmarine sandstones and shales. Producible ground water is restricted to the sandstones. Ground-water flow is northerly, toward the nearby San Juan River. The major water-bearing sandstone at the mine is the Pictured Cliffs Sandstone. The quality of ground water in this unit is poor and it is generally not tapped by wells.

Office of Surface Mining (OSM) regulations require that after strip mining, the land be restored to its original recharge capacity. This is a tall order involving a number of unknowns. First,

FIGURE 13-7 Closed depression constructed during reclamation to contain runoff at the Navajo Coal Mine, northwestern New Mexico. Auger rig is taking core for an investigation of the impact of such depressions on local ground-water recharge.

it assumes that recharge in undisturbed ground can be determined. Second, it assumes that a pre-mining value is available. Third, it assumes that recharge can be determined in disturbed ground. Fourth, it assumes that a post-mining value is known. Fifth, it assumes that the post-mining recharge capacity can then be adjusted to the pre-mining value by means of earth-moving equipment.

Perceived Impact of Reclamation Practices

Coal has been extracted from the Fruitland Formation (Cretaceous) by strip mining at the Navajo Mine in northwestern New Mexico, since 1963. The mine has a policy of zero discharge, that is, no runoff is allowed to leave the property. The mine originally accomplished this by creating depressions during the regrading phase of reclamation (Figure 13-7). The idea was that runoff would be collected in these depressions and evaporate. However, OSM was concerned that this practice disturbed the hydrologic cycle by increasing ground-water recharge. Furthermore, until site-specific data could be obtained, they required that the mine assume, for planning purposes, that 100% of the average annual precipitation (6 in) became recharge.

Recharge Study

In response to these concerns, a recharge study was conducted using a chloride mass-balance approach (Stone 1990). Impact of mining on annual recharge was assessed by comparing average pre- and post-mining values (Table 13-1). More specifically, recharge was determined by applying the chloride method to a total of 29 sites, representing three typical undisturbed landscape settings and two typical reclaimed settings.

The study produced several useful results:

1. Actual values for recharge were obtained, to replace the prescribed 100% of precipitation.
2. The average recharge rate was found to be greater in reclaimed areas (up to 0.04 in/yr) than in undisturbed ground (0.02 in/yr).
3. However, even the elevated value is so small that an impact on regional ground-water resources is unlikely.
4. The chloride mass-balance method proved to be a useful means of determining pre- and post-mining recharge.

TABLE 13-1 Results of Ground-Water-Recharge Study, Navajo Coal Mine, Northwestern New Mexico

Setting	No. of sites	Recharge range (mm/yr)	Mean[a] recharge (mm/yr)
Undistrubed	(7)	(0.05–2.30)	(1.07)
Badlands	3	0.05–0.25	0.15
Upland flat	3	0.50–1.30	0.76
Valley terrace	1	2.30	2.30
Reclaimed	(22)	(0.25–12.40)	(2.29)
Depression	9	0.76–12.40	4.06
Slope/divide	13	0.25–5.80	3.02

[a]Simple arithmetic average.

Source: Modified from Stone 1990

As a result of the study, the mining company was not required to fill in the closed depressions. However, it volunteered to discontinue the practice of constructing them in future reclamation operations.

Carlin Gold Trend, Nevada

The Carlin Trend is a northwest-trending series of deposits of submicroscopic gold in northeastern Nevada. Although mining continues by underground methods in some of the deposits, extraction was initially by open-pit methods. The ore is heap-leached with cyanide to mobilize the gold. As a basis for addressing the whole range of water problems potentially encountered in mining, a hydrogeologic study of the region was conducted (Stone and others 1991). Additionally, special studies were conducted to prepare for dewatering and to evaluate concerns shared by the mining companies, government agencies, and the general public over the potential for mining to impact both water quantity and quality in the region.

Dewatering

The mine pits are expected to extend beneath the regional water table, in some cases by as much as 1,000 ft. A series of hydrogeologic studies was required just for the dewatering aspects of the mining operation:

1. pumping tests to determine aquifer properties;
2. numerical ground-water modeling to determine the best design for the dewatering system;
3. evaluation of various options for disposing of the water produced by dewatering (reinjection, irrigation, evaporation, wetlands, and release to the Humboldt River);
4. location and evaluation of a suitable site for a reservoir to periodically store dewatering product when it can not be disposed of; and
5. chemical studies of the dewatering product to determine what treatment would be appropriate should the disposal option selected require it.

The design of the dewatering system was based largely on the results of a three-dimensional, finite-element, numerical, ground-water model (Figure 13-8). This model was constructed using

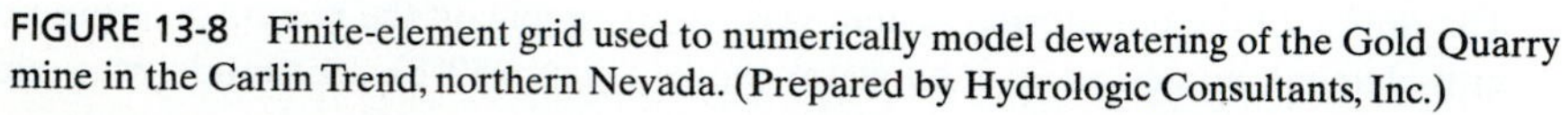

FIGURE 13-8 Finite-element grid used to numerically model dewatering of the Gold Quarry mine in the Carlin Trend, northern Nevada. (Prepared by Hydrologic Consultants, Inc.)

geologic and hydrologic data collected during mining and drilling of monitoring and production wells. Hydraulic parameters were provided by pumping tests on the various aquifers.

Impacts on Water Quantity

Quantity concerns centered around lowering of water levels as a result of dewatering. The Bureau of Land Management was concerned that such water-level declines would dry up the regional springs that serve as a water supply for both livestock and wildlife. An additional concern was that dewatering would lower the water level in the regional aquifer. This could cause the drying up of wells, changes in flow directions, and a decrease in the amount of ground-water discharge to the Humboldt River.

In response to concerns over impact on the springs, a special study was undertaken. Various types of springs were identified (Figure 6-9). Of particular interest, however, were those occurring fairly high up on the side of one of the ranges (Marys Mountain). It was concluded that these springs represent a perched artesian system within the mountain block (Stone 1993). Although this conceptual model is unusual, it is supported by several lines of evidence.

1. The springs all occur at approximately the same elevation.
2. Numerical modeling, based on the assumption that the springs are merely manifestations of a higher regional water table in the range, showed that there was no transmissivity low enough to keep water level that high, if indeed it represents the regional water table. Thus, the water discharging from the springs must represent a perched system.
3. Although drilling in the area had never extended deep enough to determine whether an unsaturated zone exists below the spring-water system, perching seems reasonable based on local geologic and hydrologic conditions. The Roberts Mountains Thrust, a large horizontal fault surface of regional extent, could be responsible. Gouge along the thrust could form the base of the perched saturation and the low-permeability fill of the adjacent basins could form the lateral boundaries.
4. Exploration drilling, at a location on the slopes of the range, had encountered flowing water that rose to what calculations showed to be the same elevation as that of the springs. This suggested that the springs may be artesian as well.
5. Confinement could be provided by both the fracturing and smaller listric faults associated with the thrust.
6. In a well drilled at one of the springs to test the conceptual model, water flowed to the surface, confirming artesian conditions.

Therefore, it was concluded that these springs represent a separate hydrologic system that should not be impacted by mine dewatering.

The impact of mining on regional water levels was studied by the USGS (Maurer and others 1996). Pumping at the two largest mines was 5,000 acft/yr in 1988, 100,000 acft/yr in 1993, and could exceed 135,000 acft/yr by the year 2000. The main impact has been the removal of large volumes of water from storage. Such losses are associated not only with pumping for dewatering and water supply, but also with increased evapotranspiration, where reservoir leakage has raised the water table, in places to the land surface. Observed water-level changes range from a decline of 800 ft at one pit to a rise of 70 ft near a leaky reservoir (Figure 13-9).

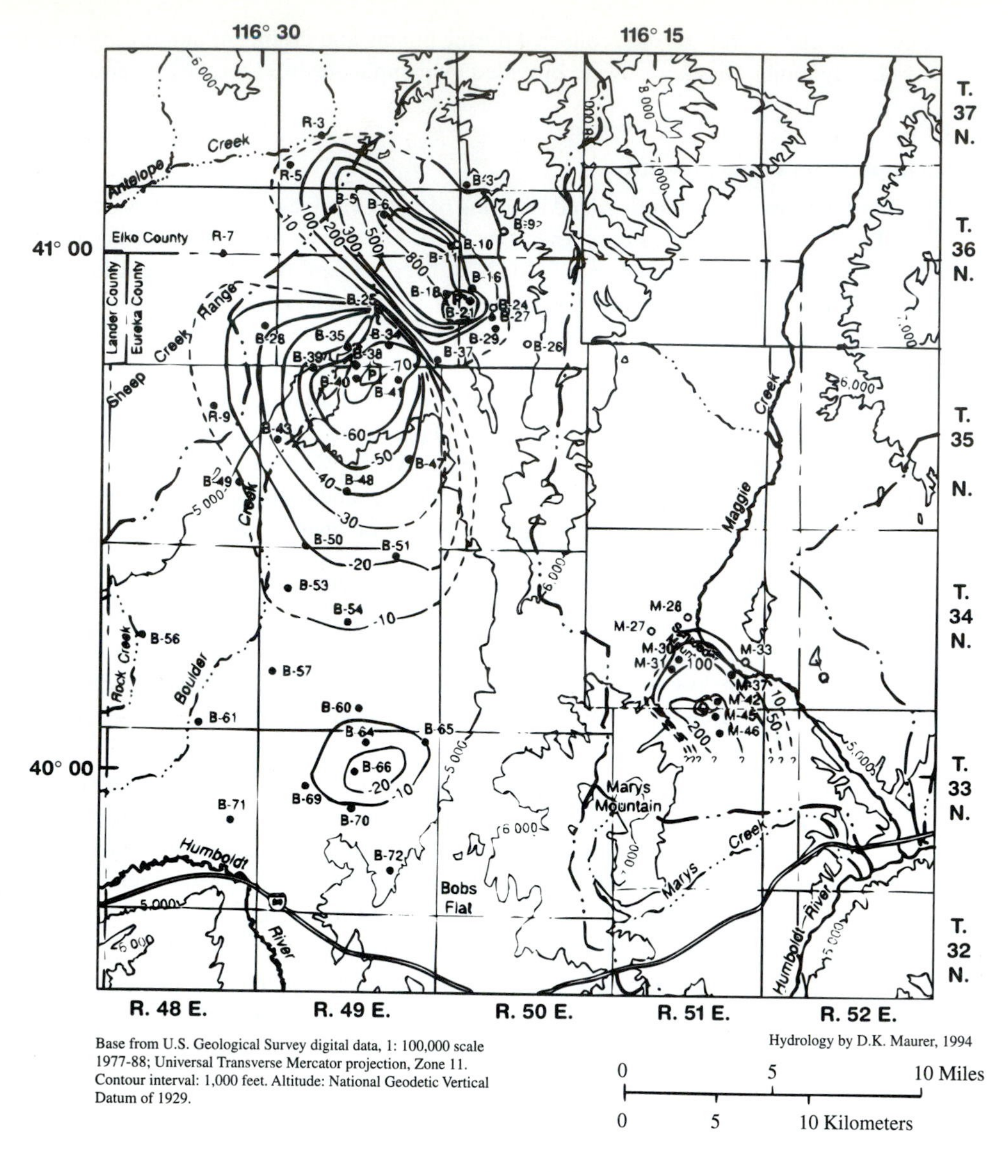

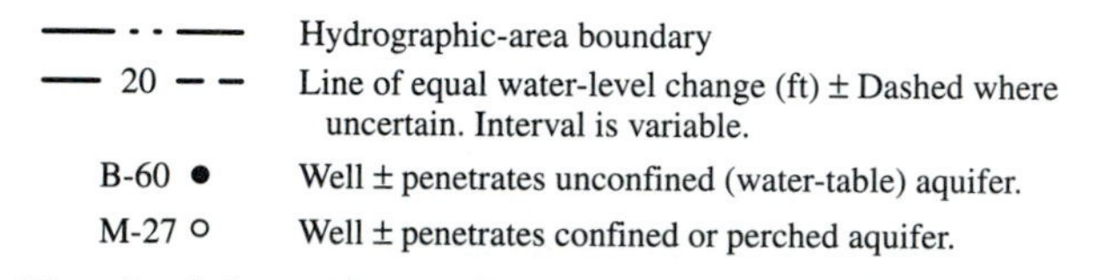

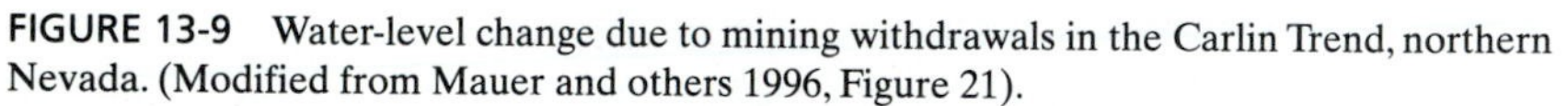
FIGURE 13-9 Water-level change due to mining withdrawals in the Carlin Trend, northern Nevada. (Modified from Mauer and others 1996, Figure 21).

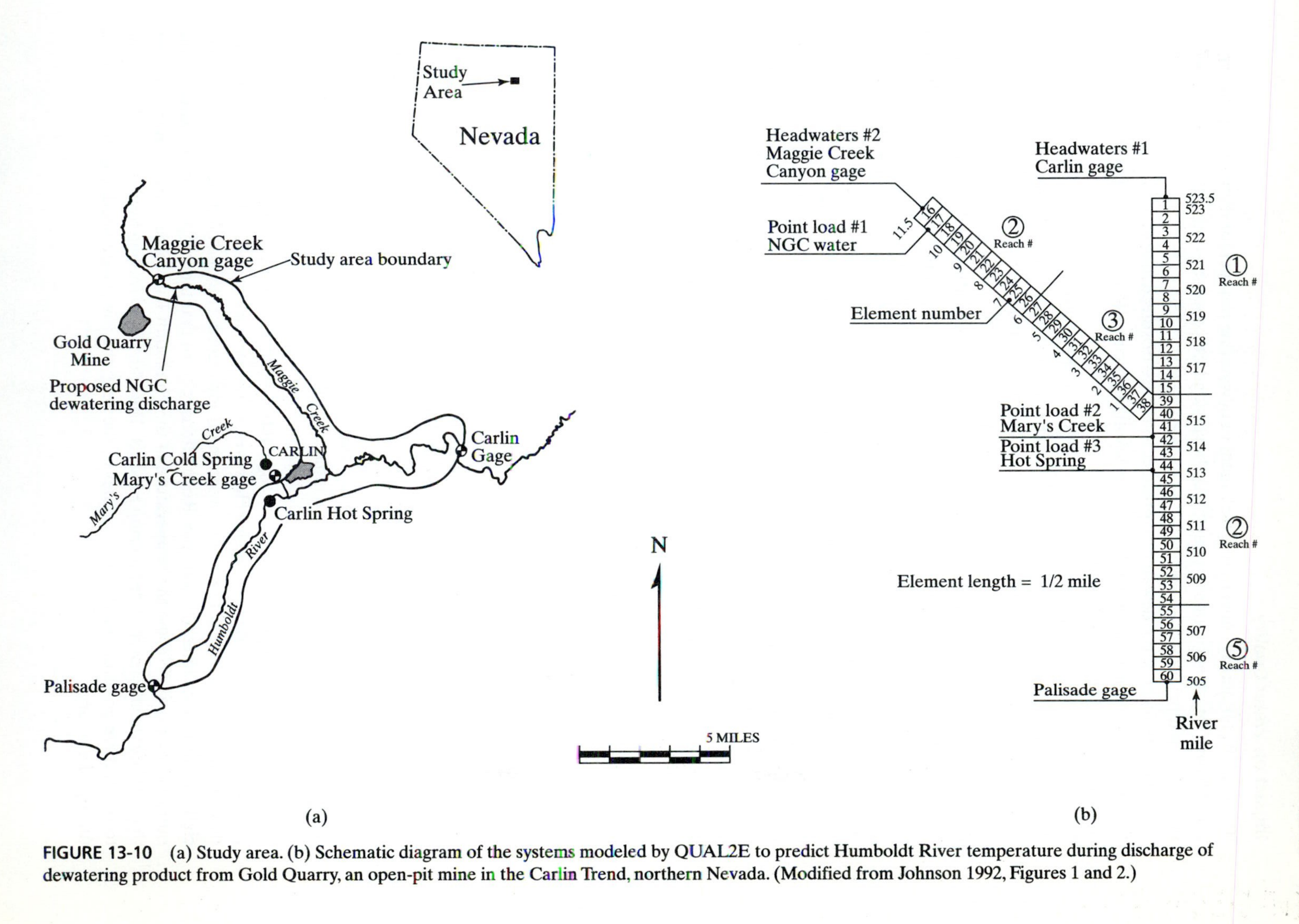

FIGURE 13-10 (a) Study area. (b) Schematic diagram of the systems modeled by QUAL2E to predict Humboldt River temperature during discharge of dewatering product from Gold Quarry, an open-pit mine in the Carlin Trend, northern Nevada. (Modified from Johnson 1992, Figures 1 and 2.)

Impact on Water Quality

The main water-quality concern was treatment of dewatering product for disposal into the Humboldt River. Pit water originally contained above-standard concentrations of arsenic and selenium and its temperature is elevated (77° at the well head).

The treatment of the dewatering product to reduce dissolved arsenic was seen as fairly routine. The conventional method at area mines involves precipitation with ferric hydroxide. A treatment plant was constructed for this purpose.

However, the removal of dissolved selenium from the dewatering product was problematic. Approximately 80% of this constituent in the pit water was found to occur in a form that makes its treatment difficult (the +6 oxidation state). Selenium in the +4 oxidation state is typically removed by adsorption onto or coprecipitation of iron oxides and hydroxides through the addition of iron salts. However, +6 selenium is more soluble and poorly adsorbed by iron oxides and hydroxides. For this species of selenium, reduction to a +4 state must precede the typical treatment.

The removal of +6 selenium from the pit water was thoroughly investigated. One study investigated the effectiveness of four iron species in removing +6 selenium from both a prepared solution and actual pit water, at various values of pH (Dubois 1992). The best results were obtained with iron filings; removal of +6 selenium improved with decreasing pH. It was concluded that bicarbonate in the pit water interfered with the reduction process. Thus, it was recommended that treatment should first involve removal of bicarbonate (either by precipitation with calcium or acidification and agitation to drive off CO_2).

However, the issue of treating dewatering product is now moot, and the results of this investigation will not be needed, at least at Gold Quarry. Soon after dewatering was initiated, arsenic and selenium content in the pit water reportedly leveled off. Apparently the volume of arsenic- and selenium-enriched ground water was relatively small and limited to the immediate vicinity of the mineralized zone of the pit. Dewatering product being pumped now originates farther away from the pit and thus is "cleaner."

Anticipating that the temperature of the dewatering product at the disposal point could not be significantly greater than that of the receiving waters, another special study was undertaken (Johnson 1992). Although the river water is warmer than would be expected, due to the presence of hot springs close to the discharge point, it was not clear whether dewatering product would cool sufficiently during storage and conveyance to be compatible with it. This problem was addressed by preliminary modeling with QUAL2E (Figure 13-10). More specifically, cooling during each of three months, covering the range of climatic conditions, was modeled: January (winter, low river flow), June (spring and early summer, high flow) and September (late summer, low flow). Separate simulations were also made for day and night cooling, except for the January scenario, since temperatures then fall below the range handled by QUAL2E.

The study showed that in the June and September scenarios, the dewatering product would be sufficiently cooled to the temperature range of the receiving water (Humboldt River) by transport along a tributary (Maggie Creek). However, in the January scenario, the river is colder and dewatering product may not be cooled enough during transport to match its temperature, especially for discharge rates >20,000 gpm. Regulations require that the temperature of water at the outfall not exceed 77°F and the temperature of the Humboldt River must not be changed more than 2°C by the discharge of dewatering product. The addition of cooling towers made it possible to reach these goals during the cooler times of the year; the towers are needed from September through February.

REFERENCES

Adrian Brown Consultants, Inc. 1991. Report on hydrology and water quality, Terrero Mine, near Terrero, San Miguel County, New Mexico. Consulting report. Adrian Brown Consultants, Inc. 75 p.

Australian Federal Environment Department. 1995. Best practice environmental management— a series of modules. 11 modules. Australian Federal Environment Department, Environment Protection Agency.

Boyles, J. M., D. Cain, W. Alley, and R. M. Klusman. 1974. Impact of Argo Tunnel acid mine drainage, Clear Creek County, Colorado. In water resources problems related to mining. *Proceedings* 18. R. F. Hadley and D. T. Snow, eds. American Water Resources Association, p. 41-53.

Camp, Dresser and McKee. Inc. 1988. Factors affecting the geochemistry of the Berkeley PIT, Butte, Montana. Consulting report. Camp, Dresser and McKee, Inc.

DuBois, M. A. 1992. Gold Quarry (GQTW-4) dewatering product bench scale water treatment—selenium removal study. Company report. Hydrology Department, Newmont Gold Co. 17 p.

Fernandez-Rubio, R., S. F. Lorca, and J. E. Arlegui. 1986. Abandono de minas impacto hidrologico. Instituto Geologico y Minero de Espana. 267 p.

Hadley, R. F., and D. T. Snow, eds. 1974. Water problems in mining. *Proceedings* 18. American Water Resources Association. 218 p.

Johnson, D. L., 1992. Prediction of Humboldt River temperatures during discharge of Gold Quarry dewatering product using QUAL2E. Company report. Hydrology Department, Newmont Gold Co. 15 p.

Kelly, T. E., R. L. Link, and M. R. Schipper. 1980. Effects of uranium mining on ground water in Ambrosia Lake area, New Mexico. *Memoir* 38: 313–19. New Mexico Bureau of Mines and Mineral Resources.

Kesseru, Z., ed. 1979. Selected papers on mine water control. *Proceedings of the Mining Development Institute.* Budapest, Hungary. 142 p.

Lyford, F. P., P. F. Frenzel, and W. J. Stone. 1980. Preliminary estimates of effects of uranium-mine dewatering on water levels, San Juan Basin, New Mexico. *Memoir* 38: 320–33. New Mexico Bureau of Mines and Mineral Resources.

Mauer, D. K., R. W. Plume, J. M. Thomas, and A. K. Johnson. 1996. Water resources and effects of changes in ground-water use along the Carlin trend, northern-central Nevada. Water-resources investigations report 96-4134. U.S. Geological Survey. 146 p.

McGurk, B. E., and W. J. Stone. 1986. Conceptual hydrogeologic model of the Nations Draw area, Catron and Cibola County, New Mexico. Report to Salt River Project. New Mexico Bureau of Mines and Mineral Resources. Phoenix. 104 p.

New Mexico Energy, Minerals, and Natural Resources Department. 1990. State permit requirements for development of energy and mineral resources in New Mexico. Report of Investigation no. 5. New Mexico Energy, Minerals. and Natural Resources Department, Mining and Minerals Division, Mineral Industry Services. 71 p.

Stone, W. J. 1984. Recharge in the Salt Lake coal field based on chloride in the unsaturated zone. Open-file report 214. New Mexico Bureau of Mines and Mineral Resources. 64 p.

Stone, W. J. 1990. Impact of mining on ground water recharge, Navajo coal mine, New Mexico. *Mining Engineering* 42 (11): 11–14.

Stone, W. J. 1993. Perched artesian conditions in mountainous terrain (abs). *Abstracts and Program.* Rocky Mountain Ground Water Conference.

Stone, W. J., T. Leeds, R. C. Tunney, G. A. Cusack, and S. A. Skidmore. 1991. Hydrology of the Carlin Trend, northeastern Nevada—a preliminary report. Company report. Hydrology Department, Newmont Gold Co. 123 p.

Stone, W. J., F. P. Lyford, P. F. Frenzel, N. H. Mizell, and E. T. Padgett. 1983. Hydrogeology and water resources of San Juan Basin, New Mexico. Hydrologic report 6. New Mexico Bureau of Mines and Mineral Resources. 70 p.

CHAPTER 14

Ground-Water Modeling

Hydrologic studies increasingly involve ground-water modeling. Hydrogeologists must, therefore, be at least model-literate, if not model-proficient. This brief overview is intended to be a starting place for hydroscientists who may be unfamiliar with the topic and a review or clarification for those with some knowledge or experience with models. Detailed coverage of the subject is beyond the scope of this book. There are numerous references on hydrologic modeling, some of which are listed at the end of the chapter.

A model may be defined as an idealized representation of reality, used to demonstrate and illustrate some aspect of it. In the context of hydrologic models, "reality" is the hydrogeologic system and "some aspect of it" depends on the objective of the study. Modeling has been applied to surface-water, soil-water, and ground-water problems alike. Although the discussion here emphasizes ground-water models, the main tasks and procedures apply equally well to models of the other parts of the hydrologic cycle.

TYPES

There are many kinds of ground-water models. They differ in terms of the process and condition addressed as well as the geometry and approach used.

By Process

Ground-water models may address flow, transport, or deformation conditions. Flow models focus on ground-water movement. They are useful in studying water-supply potential, regional aquifers, well performance, stream and aquifer interaction, injection programs, and dewatering scenarios. Transport models generally deal with the aqueous movement of solutes or heat. Solute-transport models are applied to waste-disposal, contaminant-remediation, and salt-water–intrusion problems. Ground-water heat-transport models are used in geothermal reservoir analysis. There are also heat-transport models for surface-water systems. These models are applied to problems of thermal pollution of streams. Deformation models help characterize or anticipate subsidence due to fluid withdrawal.

By Condition

Ground-water models may be applied to either of two conditions: steady-state or transient. The steady-state condition is the first to be simulated in any ground-water modeling project. It represents the natural or equilibrium condition of the hydrologic system. By contrast, the transient or disequilibrium condition is that resulting from a specific stress (such as pumping, injection or leakage). Models of transient conditions are used to understand or estimate the response of the hydrologic system to a given stress.

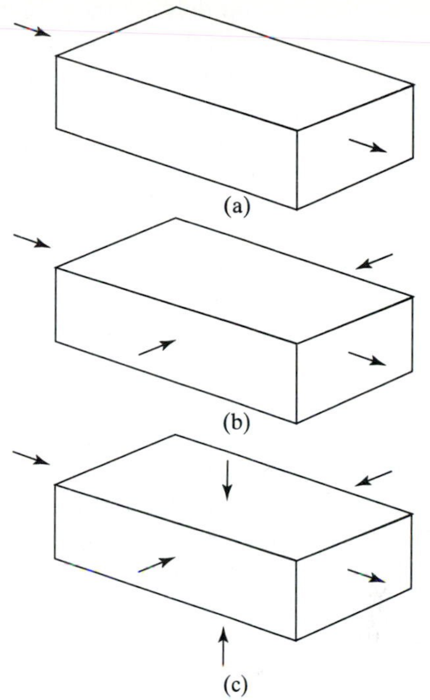

FIGURE 14-1 Schematic diagram of ground-water model geometries: (a) 1-D. (b) 2-D, (c) 3-D. Arrows indicate ground-water flow directions addressed.

By Geometry

The geometry of a ground-water model depends on how the hydrologic system is conceptualized; it may be one-, two-, or three-dimensional (Figure 14-1). A one-dimensional (1-D) model may be used to evaluate water balance (for example, Stone 1995). A two-dimensional (2-D) geometry may be sufficient where there is mainly horizontal flow (for example, O'Brien and Stone 1984). Three-dimensional (3-D) models are the most complex, accounting for both horizontal and vertical ground-water movement (for example, Lyford and others 1980).

By Approach

Ground-water models also vary as to the approach taken: physical, analog, or mathematical. Physical models include such things as sand boxes, flumes, or columns. Electrical analog models were the first nonphysical types used. In these, ground-water movement was simulated as the flow of electrical current. Modern ground-water models are generally mathematical. A mathematical model is a set of equations describing the processes in an aquifer, or group of aquifers, for the assumptions made.

Mathematical models are based on a physical system, and thus are governed by physical laws and concepts. These models make simplifying assumptions that simplify the model and put physics (conservation of mass and momentum) into mathematical terms, to form a governing equation. Mathematical models may be stochastic or deterministic and analytical or numerical. Some combine aspects of more than one of these forms. Stochastic models are those in which there are uncertainties in the values of the output variables (i.e., they are random variables). Deterministic models do not involve randomness; there is no statistical uncertainty in the values of the output variables. Analytical models simplify equations so they can be solved analytically (e.g., the Theis curve). Numerical models replace the continuous differential equations with a finite number of algebraic equations that can be solved by matrix techniques with a computer. More specifically, the continuous variables are replaced by discrete variables, defined at nodes in the model grid. These variables are discrete for both space and time. Numerical models can be applied to more complicated and realistic systems than can analytical models.

Two main methods are used: finite-difference and finite-element. They are easily distinguished by their grid style (Figure 14-2). The grid for finite-difference models is square or rectangular. That is, grid cells or blocks are formed by two sets of lines intersecting at right angles to define columns and rows. Data input and output points or nodes may be located at either the center or the corners of the cells. By contrast, the grid for finite-element models consists of a mesh of squares, rectangles, triangles, or polygons. Nodes are located at the corners of these elements. For 3-D models, the same grid applies to all layers, regardless of method.

There are also mathematical differences between the two methods. For example, in the finite-difference method, equations are solved by differentiation, whereas in the finite-element method solution is by integration. Also, in the finite-difference method, output applies to the node, but is assumed to be constant throughout the cell with which it is associated. In the finite-element method, output may be determined (interpolated) for any individual point between nodes, within the area covered by the grid.

CONSTRUCTION

Setting up and using a ground-water model involves various tasks (Figure 14-3):

defining the problem,
determining whether a model is needed,
formulating the conceptual model,
developing the steady-state numerical model,
calibrating the steady-state model,
developing the transient model, and
calibrating the transient model.

A model is never done; rather it is constantly modified as additional information becomes available.

Defining the Problem or Need for Modeling

The first step in any modeling project is to identify the problem to be solved. What hydrologic property is to be learned? What hydrologic question is to be answered? Then, before going further, determine whether a model is really necessary. Can the information

FIGURE 14-2 Comparison of finite-difference and finite-element grids: (a) hypothetical aquifer, (b) possible block-centered finite-difference grid, and (c) possible finite-element grid. (From Mercer and Faust 1980, Figure 4. Reprinted by permission of Ground Water Publishing.)

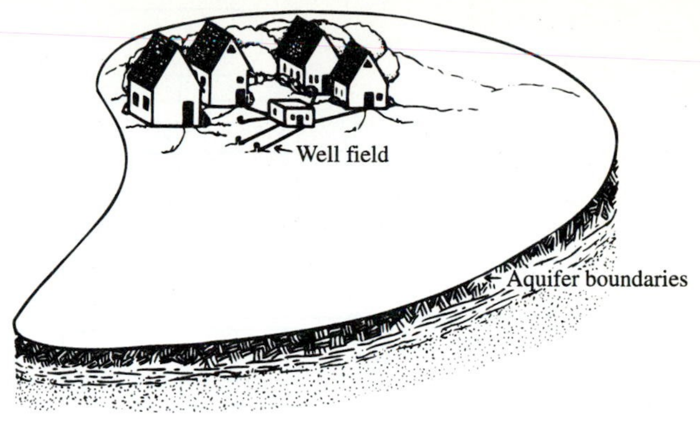

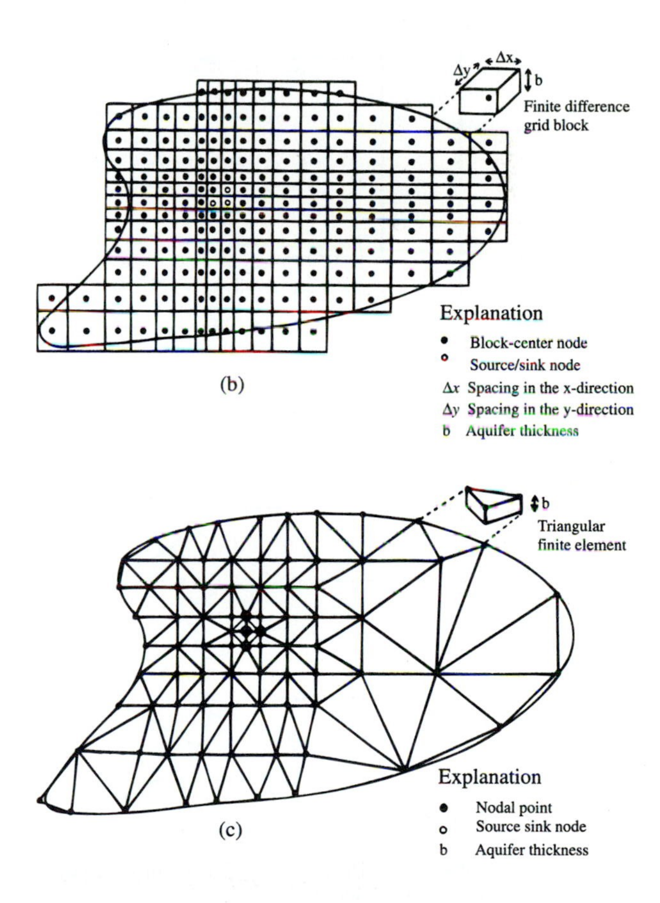

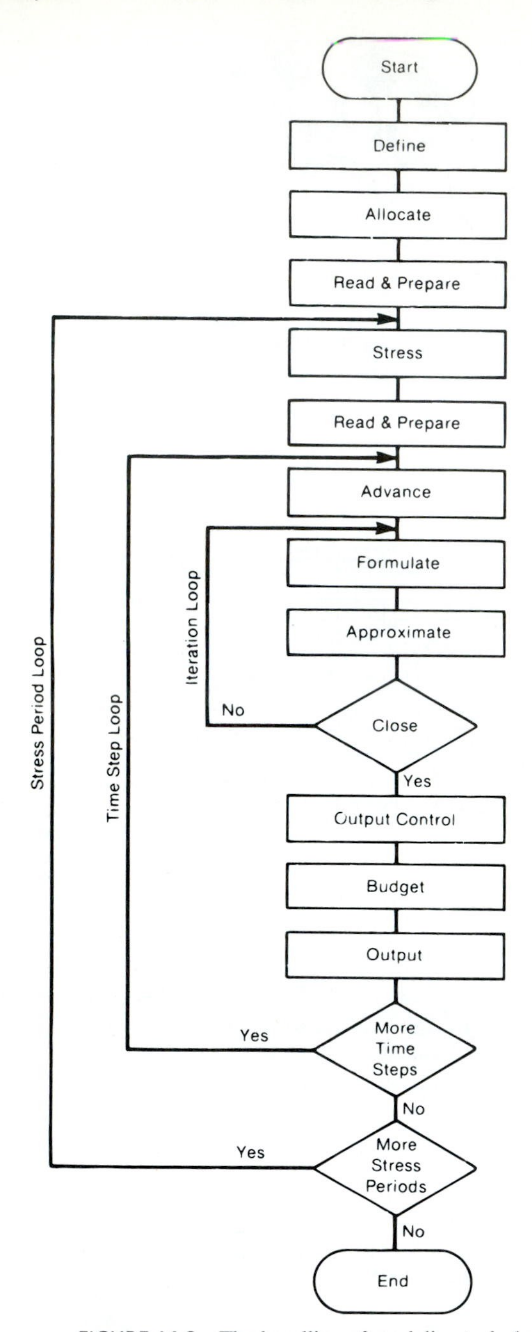

DEFINE — Read data specifying number of rows, columns, layers, stress periods, and major program options.

ALLOCATE — Allocate space in the computer to store data.

READ AND PREPARE — Read data which is constant throughout the simulation. Prepare the data by performing whatever calculations can be made at this stage.

STRESS — Determine the length of a stress period and calculate terms to divide stress periods into time steps.

READ AND PREPARE — Read data which changes from one stress period to the next. Prepare the data by performing whatever calculations can be made at this stage.

ADVANCE — Calculate length of time step and set heads at beginning of a new time step equal to heads calculated for the end of the previous time step.

FORMULATE — Calculate the coefficients of the finite difference equations for each cell.

APROXIMATE — Make one cut at approximating a solution to the system of finite difference equations.

OUTPUT CONTROL — Determine whether results should be written or saved on disk for this time step. Send signals to the BUDGET and OUTPUT procedures to indicate exactly what information should be put out.

BUDGET — Calculate terms for the overall volumetric budget and calculate and save cell-by-cell flow terms for each component of flow.

OUTPUT — Print and save heads, drawdown and overall volumetric budgets in accordance with signals from OUTPUT CONTROL procedure.

FIGURE 14-3 The handling of modeling tasks by MODFLOW. (From McDonald and Harbaugh 1988, Figure 13.)

needed be acquired by simpler methods? If a model is, indeed, required, the hydrologic question to be answered must be turned into a problem the modeling can address.

Formulating a Conceptual Model

Before constructing a numerical model can be constructed, formulate a conceptual model of the hydrogeologic system. As discussed in chapter 8, this requires a compilation and interpretation of existing information on both the geology and hydrology of the study area. This background will not only provide an understanding of the physical system to be modeled, but will also identify the need for additional data. This step is important since the conceptual model provides the assumptions for the numerical model. Thus, the numerical model will only be as good as the conceptual model upon which it is based. Initial runs of the model may suggest the conceptualization of the system is not adequate and modifications must be made.

Developing the Mathematical Model

Once the hydrogeologic system has been conceptualized, construct a mathematical model. This involves several tasks:

selecting a model code,
discretizing space and time,
specifying boundary and initial conditions, and
defining aquifer parameters.

These tasks do not occur in order. They may overlap and be iterative.

The code selected depends on the complexity and geometry of the system to be modeled and the capability of the computer available. Discretization of space depends on the model type: finite-difference or finite-element. In both cases, decide on the number of layers to be used. If it is a finite-difference model (Figure 14-4), not only the number and size of rows and columns must be chosen, but also whether the grid is to be block-centered (nodes in the center of blocks) or mesh-centered (nodes at block corners). If the model is a finite-element one, determine the number, size, and shape of grid elements. Time must also be discretized. Select the number and length of time steps and stress periods in the model. Boundary conditions, or those surrounding the grid, must be designated. These conditions may be set as specified-head, constant-head, specified-flux, head-dependent (general-head), or no-flow. Initial conditions consist of known water levels or heads for steady-state modeling and the steady-state conditions for transient simulations. Aquifer parameters include hydraulic conductivity, saturated thickness, storage coefficient, and recharge (or precipitation and evapotranspiration). In transient simulations, the rate of stress (pumping, injection, etc.) must also be defined.

IMPLEMENTATION

After hydrologic and mathematical parameters have been defined, run the model. A steady-state model is established before any transient simulations can be run. To do this,

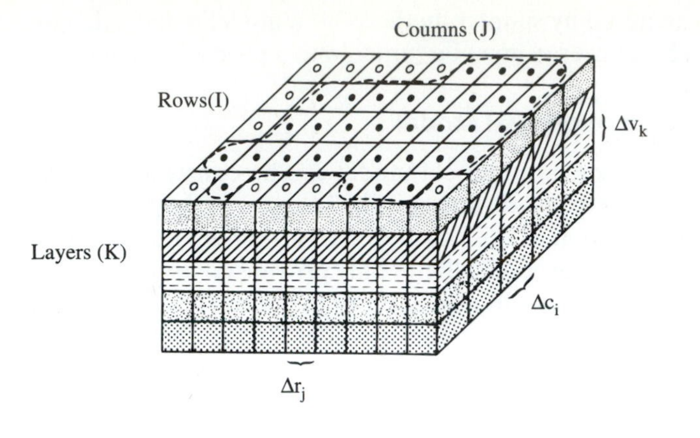

– – – –	Aquifer boundary
•	Active cell
o	Inactive cell
Δr_j	Dimension of cell along the row direction.Subscript (j) indicates the number of the column.
Δv_k	Dimension of cell along the column dircetion. Subscript (i) indicates the number of the row.
Δc_i	Dimension of cell along the vertical direction. Subscript (k) indicates the number of the layer.

FIGURE 14-4 Discretization of a hypothetical aquifer system into layers, rows and columns for 3-D modeling. (From McDonald and Harbaugh 1988.)

run the model for the best approximation of equilibrium conditions possible. When the steady-state model provides results consistent with stated acceptance criteria, evaluate it further by comparing the modeled heads (water levels) with observed heads. Adjust aquifer parameters until the comparison is as close as possible, without making unreasonable changes. When the match is good, the model is said to be *calibrated.* Now the sensitivity analysis may be carried out; that is, input parameters are systematically varied and the impact on results observed and reported.

Next, apply stress to the steady-state system, and develop a transient model. Calibration involves "history matching." This means matching the historical and modeled responses to a historical stress. Once a transient model is calibrated, it may be used to estimate the response of the system to other stresses.

LIMITATIONS

Although models are intended to address reality, they can only approximate it. Therefore, it is important to be aware of their limitations.

Verification and Validation

Reports on modeling studies often include claims about verification or validation of the model. *Verification* may be defined as demonstrating the ability of a generic model to solve the governing equation. By contrast, *validation* may be defined as demonstrating the ability of a site-specific model to represent cause-and-effect relationships. Both imply authentication of the truth and accuracy of the model, that is, that the model represents reality. Unfortunately, the correctness of a model cannot be proven, only disproven. As frankly presented by Konikow and Bredehoeft (1992), a good history match does not constitute validation or signify a predictive tool, and the terms *verification* and *validation* have no place in hydrology.

Value of Models

Many serious public-policy issues (for example, global climate change, nuclear-waste disposal, etc.) are addressed by models. However, since models provide nonunique solutions, limited to the code used and assumptions made, and since models cannot be verified or validated, you might reasonably ask, "What good are they?" In spite of their shortcomings, models are useful tools for critical analysis in hydrology. Because they are run on computers, models permit us to deal with large, complex natural systems. Additionally, they allow us to organize our thinking, test hypotheses, evaluate the sensitivity of a system to various parameters, and identify data needs or topics for further study.

Pitfalls

Models have a bad reputation among the scientific and lay community alike because they are frequently misused. Four common pitfalls of model use have been recognized (Mercer and Faust 1980; Reddi 1990):

1. overkill
2. misconceptualization
3. inadequate parameter definition
4. blind faith

An awareness of these pitfalls should reduce their occurrence and perhaps improve the negative image of model studies.

Overkill means employing a more sophisticated model than appropriate for the problem and data at hand. Just because the system is complex, doesn't mean the model has to be. It is best to start out with a simple model. In fact, a simple, pilot model is sometimes run prior to constructing the final model.

Misconceptualization includes misunderstanding the hydrogeologic system involved, misdefining the problem, or solving it by inappropriate means. Obviously, modeling the wrong system or problem will give erroneous results. An approximate solution to an accurate problem is more appropriate than an accurate solution to an approximate problem.

Models often fail because, even though the system and problem are correctly defined, input parameters are inadequate. Unfortunately, adequate data coverage of the study area is rarely encountered. Thus, some parameters must be estimated. This process should be based on what is known of the system, previous experience, and good science. Postaudits of predictive models, such as that by Konikow (1986), are very instructive for determining the distribution and significance of data deficiencies.

The most common misuse of models is to accept them blindly. This means believing unrealistic results simply because a computer or model generated them. However, it also includes misinterpreting or overextending the results. Formulating a sound conceptual model and setting appropriate acceptance criteria may eliminate this pitfall.

Numerous ground-water modeling programs or codes already exist. Many of these are in the public domain and are free. Most hydrogeologists use these codes in their modeling projects. The International Ground Water Modeling Center is a clearing house of information on and distributor for available software. They maintain offices in both the United States and Europe: the Colorado School of Mines, Golden, CO (FAX 303 273-3278) and the TNO Institute of Applied Geoscience, The Netherlands (FAX 31 15-564800). Free model software is available on the Internet at http://h2o.usgs.gov/software/.

CASE HISTORIES

A good way to learn about modeling is to study existing models, provided the reports on them clearly document what was done. The following brief case histories provide examples of 1-D, 2-D, and 3-D ground-water models. In each, notice how the code was adapted to the problem at hand.

Mortandad Canyon, New Mexico (1-D)

Mortandad Canyon is one of numerous ephemeral watercourses dissecting the Pajarito Plateau, an expanse of Pleistocene volcanic deposits (mainly tuff), situated west of the Rio Grande, 40 mi northwest of Santa Fe, New Mexico. The canyon is cut into the tuff and floored with Quaternary alluvium derived from it. The climate is semiarid with an average annual precipitation of 19 in. Evapotranspiration is assumed to be 17 in., based on the vegetation present. Although the regional water table is very deep, the alluvium is saturated in places due to the perching of water above the less permeable tuff. Mortandad Canyon receives effluent from Los Alamos National Laboratory's liquid-radioactive–waste-treatment facility. As neither streamflow nor the shallow perched ground-water body normally extend to the lab boundary, the fate of the effluent is a concern.

A 1-D, finite-difference, pilot model, using MODFLOW (McDonald and Harbaugh 1988), was constructed to investigate water balance in a 2-mi reach of the canyon below the outfall (Stone 1995). Although the system was modeled very simply as 1 layer, 1 column, and 30 rows (Figure 14-5), hydrologic parameters were scaled for actual dimensions of the canyon for realism. Since water made available to the canyon through precipitation does not normally leave as streamflow or underflow, it must be lost through evapotranspiration (ET), downward leakage into the tuff, or some combination of the two. To evaluate this, the model was set up so that ET was maximized to a reasonable value and no downward leakage at the bottom of the model or underflow through the alluvium were allowed. The model focused on the natural system before effluent was discharged. It was reasoned that if there were leakage into the tuff without the effluent, there surely would be with it. Since the stream does not flow in the lower reaches and underflow has not been observed at the property line, the 0.2 cfs that the model showed to be leaving as streamflow must represent downward leakage into the tuff (Figure 14-6).

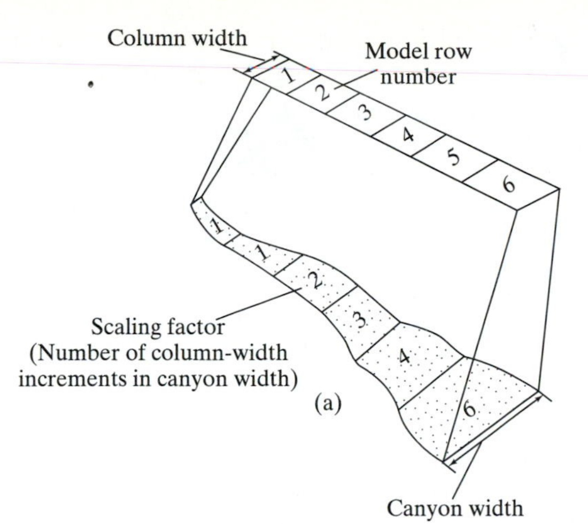

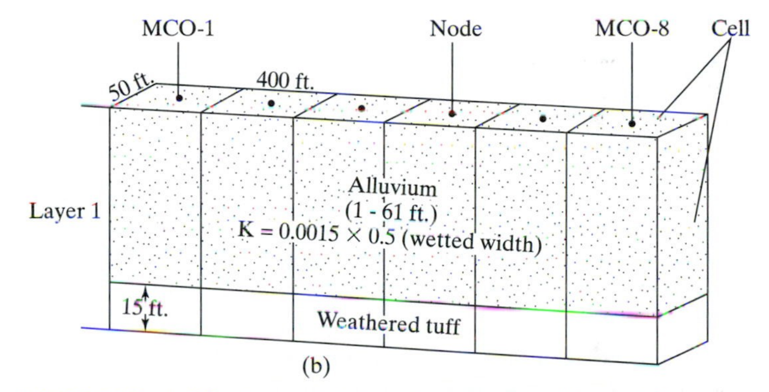

FIGURE 14-5 A 1-D pilot model of the shallow aquifer, Mortandad Canyon, Los Alamos National Laboratory, New Mexico: (a) scaling hydraulic parameters using canyon width, and (b) schematic diagram of the model. (From Stone 1995, Figures 7 and 9.)

Animas Valley, New Mexico (2-D)

The Animas Valley is an elongate, topographically closed basin in southwestern New Mexico. Geologically, the area is typical of basin-and-range country. The valley was downdropped and adjacent mountains were uplifted during Tertiary time. The mountains consist of igneous and sedimentary rocks of various ages, ranging from Precambrian to Tertiary. The valley contains Quaternary basin-fill sediments derived from the bordering ranges. Seismic studies indicate that the maximum thickness of these deposits exceeds 2,000 ft. The valley was occupied by a pluvial lake in late Pleistocene/Holocene time (Fleischhauer and Stone 1982).

The area is drained by Animas Creek, which is perennial near its origin in the mountains at the south end of the valley, but is ephemeral along most of its course. The climate straddles the arid/semiarid boundary with an average annual precipitation of 10 in. Ground water is unconfined and flow is generally toward the basin center, then northerly down the valley. Withdrawal

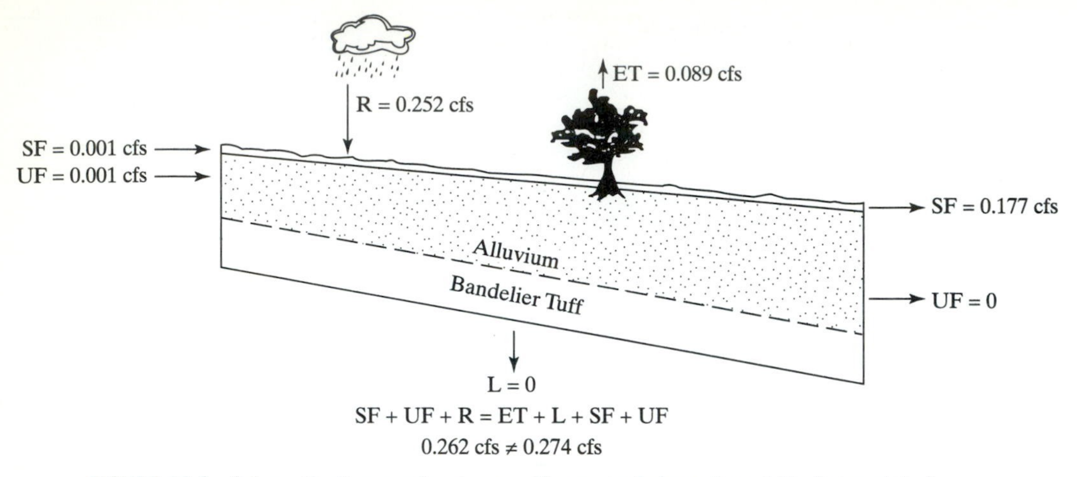

FIGURE 14-6 Schematic diagram showing resulting water balance for a 1-D pilot model of the shallow aquifer, Mortandad Canyon, Los Alamos National Laboratory, New Mexico. (From Stone 1995, Figure 10.)

of ground water for irrigation has led to water-level declines. Because the valley is typical of those in the American Southwest, it was included in the USGS's Southwest Alluvial Basin Regional Aquifer Systems Analysis.

The 2-D finite-difference model developed by Trescott and others (1976) was applied because vertical flow is considered to be insignificant in the valley. A block-centered grid was constructed using 1-mi-square cells. Boundary conditions were defined as no-flow, constant-head, or constant-flux (Figure 14-7). Transmissivity was assigned to nodes based on pumping-test results, specific-capacity data, or estimates of hydraulic conductivity from well logs and aquifer thickness from gravity-anomaly maps. After the steady-state model was calibrated, a transient model was constructed for pumping from April 1949 to April 1984 (Figure 14-8).

Sensitivity of the model to transmissivity (T), storativity (S), discharge (Q), and recharge (R) was evaluated by perturbing each plus or minus 50%. T and S were perturbed at all nodes, whereas Q and R were perturbed only at nodes where these parameters were specified. Head was found to be most sensitive to a 50% decrease in S. The study also showed the value of using all available geological and geophysical data in estimating hydrologic parameters (O'Brien and Stone 1984).

San Juan Basin, New Mexico (3-D)

The San Juan Basin is a structural depression at the eastern edge of the Colorado Plateau, lying mostly in northwestern New Mexico. Rocks of Precambrian and Paleozoic age occur in uplifts bordering the basin. Sedimentary rocks of Mesozoic and Cenozoic age crop out around the margin and across the basin. Maximum stratigraphic thickness exceeds 14,000 ft. Climate of the basin is arid, with average annual precipitation ranging from 6 to 10 in.; by contrast, as much as 30 in. fall in the adjacent mountain recharge areas.

Extensive dewatering of a major artesian sandstone aquifer in conjunction with underground uranium mining along the southern margin of the basin prompted a regional hydrogeologic study

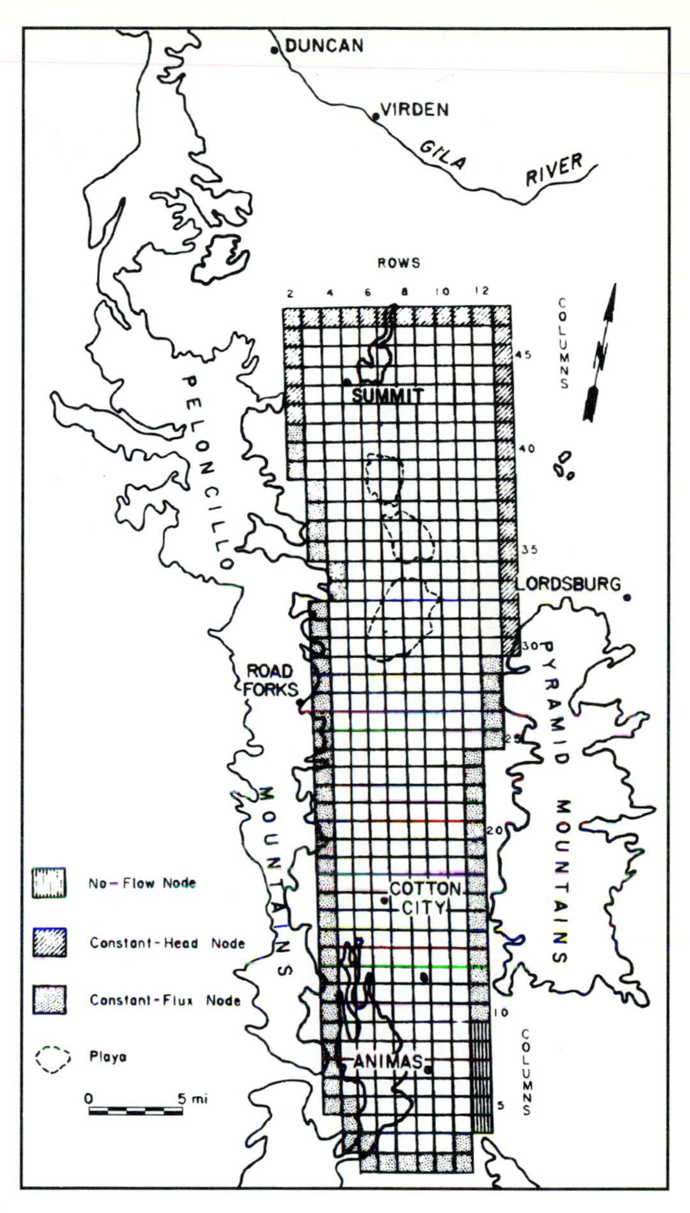

FIGURE 14-7 Grid and boundary conditions used for a 2-D model of the Animas Valley, New Mexico. (From O'Brien and Stone 1984, Figure 8. Reprinted by permission of Ground Water Publishing.)

(Stone and others 1983), as well as numerical modeling to evaluate the impact of the dewatering on the flow of the San Juan River and Rio Grande, the discharge areas for the basin (Lyford and others 1980). A 3-D, finite-difference model was constructed, using the code developed by Trescott (1975) and Trescott and Larson (1976). The system was discretized into 34 columns, 64 rows, and 4 layers (Figure 14-9). At the time, it was one of the largest ground-water models ever attempted.

The modeling addressed the impact of dewatering on the potentiometric surface associated with the aquifer for the year 2000 under three different scenarios of mining intensity. The model predicted that drawdowns near the deepest mines would be 2,000 ft for the lowest level of activ-

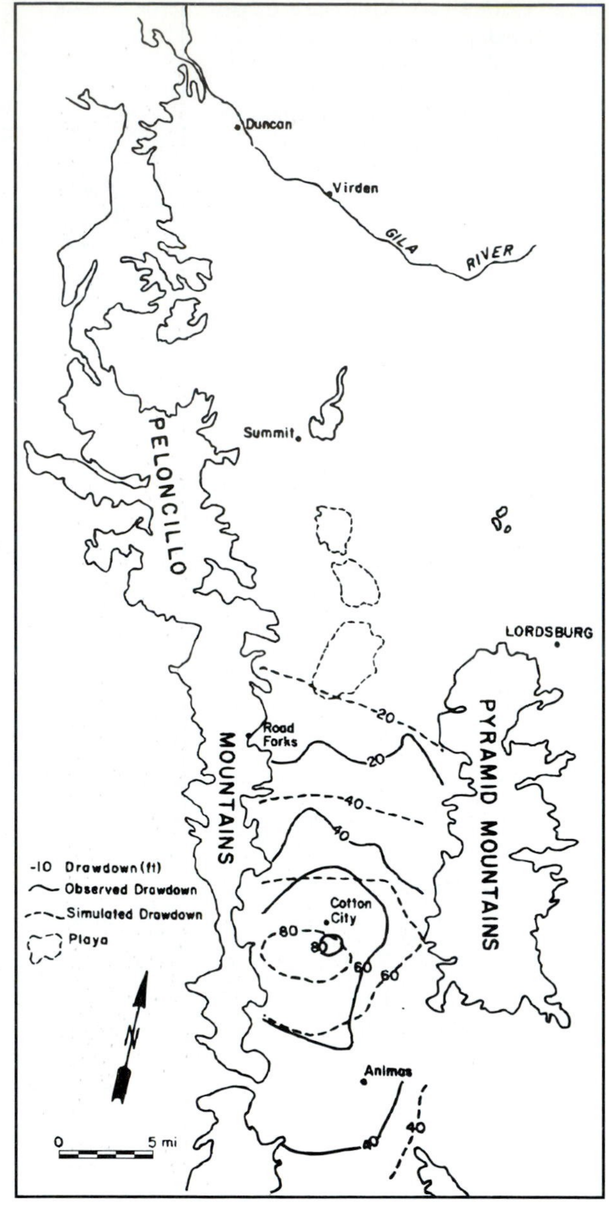

FIGURE 14-8 Results of transient run (April 1948–April 1981) for a 2-D model of the Animas Valley, New Mexico. (From O'Brien and Stone 1984, Figure 11. Reprinted by permission of Ground Water Publishing.)

ity (33 mines) and 4,000 ft for the medium (75 mines) and high (105 mines) levels of activity (Figure 14-10). Additionally, the model predicted that mining at high-activity level would result in a 0.04 cfs reduction in the flow of the San Juan River long after the year 2000 and a 0.5 cfs reduction in ground-water flow toward the Rio Grande. Sensitivity analysis was not attempted. Historical mine discharge data were scarce, but where history matching was possible, model predictions were quite good. The greatest uncertainty in the project was the number, location, and water production of mines assumed for the different scenarios of mining activity. Although the predictions

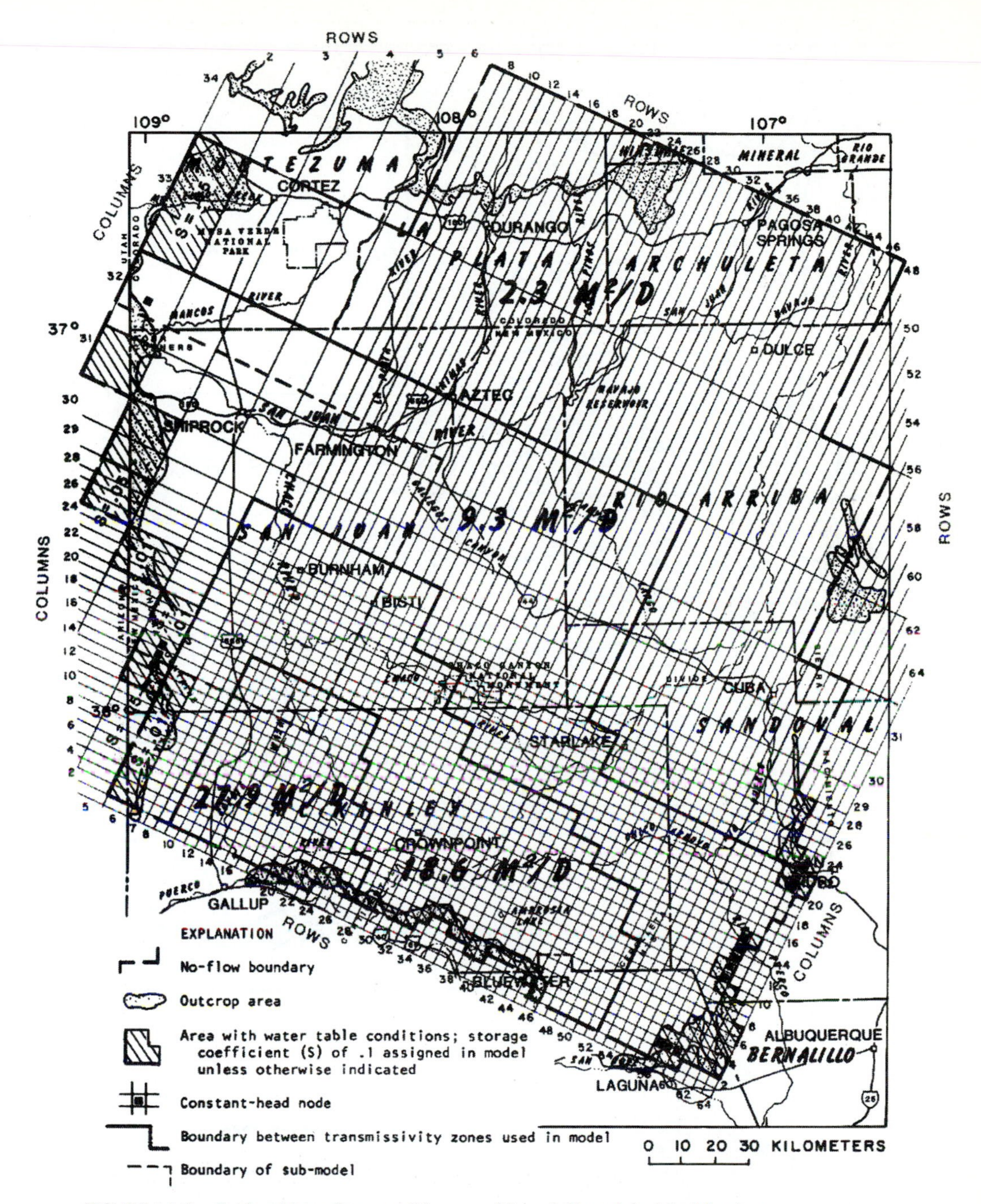

FIGURE 14-9 Grid and boundary conditions used for a 3-D model of the Morrison Formation, San Juan Basin, New Mexico. (From Lyford and others 1980, Figure 3.)

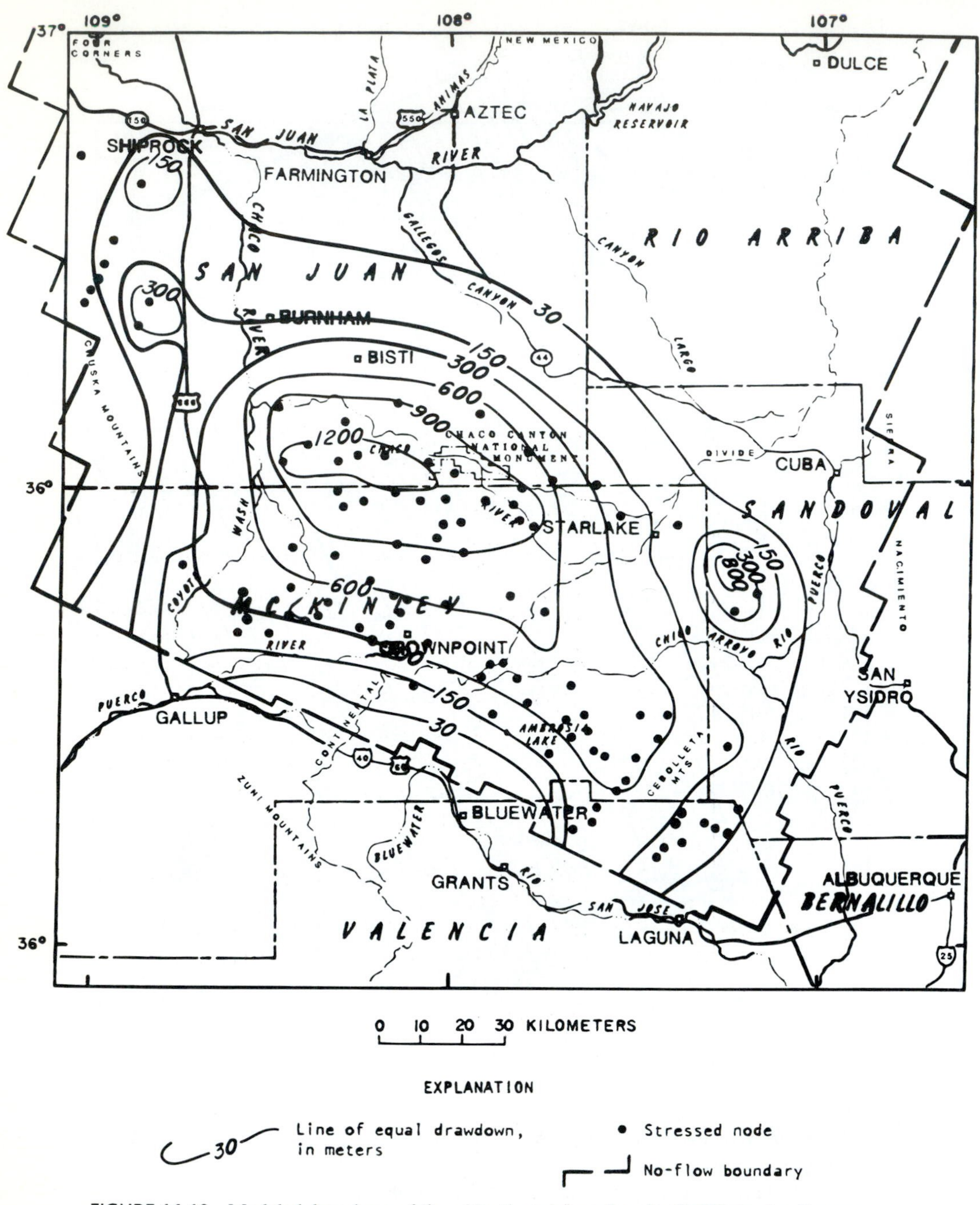

FIGURE 14-10 Modeled drawdown of the potentiometric surface for the Westwater Canyon Member, Morrison Formation, San Juan Basin, New Mexico, by the year 2000, assuming a high level of mining and dewatering activity. (From Lyford and others 1980, Figure 9.)

will never be tested due to the decline of mining activity in the late 1980s, it is an interesting application of ground-water modeling.

Used correctly, models are valuable. However, as noted by Grodin and others (1990), models should be used to complement skills, not replace them. The USGS suggests one should be a hydrogeologist first and a modeler second.

REFERENCES

Anderson, M. P., D. S. Ward, E. G. Lappala, and T. A. Prickett. 1993. Computer models for subsurface water. In *Handbook of hydrology,* p. 22.1–22.34. D. Maidmont, ed. New York: McGraw-Hill.

Anderson, M. P., and W. W. Woessner. 1992. Applied groundwater modeling. San Diego, CA: Academic Press. 381 p.

Fleischhauer, H. L., Jr., and W. J. Stone. 1982. Quaternary geology of Lake Animas, Hidalgo County, New Mexico. Circular 174. New Mexico Bureau of Mines and Mineral Resources. 25 p.

Grondin, G. H., M. Gannett, P. K. M. van der Heijde, and R. O. Patt. 1990. Critical errors that hydrogeological professionals can make with computer programs. *Proceedings,* p. 149–58. American Water Resources Association Symposium on Transferring Models to Users, Denver, CO.

Konikow, L. F. 1986. Predictive accuracy of a ground-water model—lessons from a postaudit. *Ground Water* 24 (2): 173–84.

Konikow, L. F., and J. D. Bredehoeft. 1992. Ground-water models cannot be validated. *Advances in Water Resources,* 15: 75–83.

Lyford, F. P., P. F. Frenzel, and W. J. Stone. 1980. Preliminary estimates of the effects of uranium-mine dewatering on water levels in the San Juan Basin, New Mexico. *Memoir* 38: 320–32. New Mexico Bureau of Mines and Mineral Resources.

McDonald, M. G., and A. W. Harbaugh. 1988. A modular three-dimensional finite-difference ground-water flow model. In *Techniques of Water Resources Investigations*, Book 6, Chapter A1. U.S. Geological Survey.

Mercer, J. W., and C. R. Faust. 1980. Ground-water modeling—mathematical models. *Ground Water* 18(3): 212–27.

O'Brien, K. M., and W. J. Stone. 1984. Role of geological and geophysical data in modeling a Southwestern alluvial basin. *Ground Water* 22 (6): 717–27.

Reddi, L. N. 1990. Potential pitfalls in using groundwater models. *Proceedings,* p. 131–140. American Water Resources Association Symposium on Transferring Models to Users, Denver, CO.

Stone, W. J. 1995. Preliminary results of modeling the shallow aquifer, Mortandad Canyon, Los Alamos National Laboratory, New Mexico. Report NMED/DOE/AIP-95-1. New Mexico Environment Department. 32 p.

Stone, W. J., F. P. Lyford, P. F. Frenzel, N. H. Mizell, and E. T. Padgett. 1983. Hydrogeology and water resources of the San Juan Basin, New Mexico. New Mexico Bureau of Mines and Mineral Resources. Hydrologic Report 6. 70p.

Trescott, P. C. 1975. Documentation of finite-difference model for simulation of three-dimensional ground-water flow. Open-file report 75-438. U.S. Geological Survey. 32 p.

Trescott, P. C., G. F. Pinder, and S. P. Larson. 1976. Finite-difference model for aquifer simulation in two dimensions with numerical experiments. In *Techniques of Water Resources Investigations,* Book 7. U.S. Geological Survey. 116 p.

Trescott, P. C., and S. P. Larson. 1976. Supplement to open-file report 75-438, documentation of finite-difference model for simulation of three-dimensional ground-water flow. Open-file report 76-591. U.S. Geological Survey.

Wang, H. F., and M. P. Anderson. 1995. Introduction to groundwater modeling—finite-difference and finite-element methods. San Diego, CA: Academic Press. 237 p.

CHAPTER 15

Final Suggestions

Hydrogeologic studies are vitally important today and will become more so in the future. Thus, it is a disservice not only to employers but to the public to do them poorly. The preceding chapters have shown what constitutes a sound hydrogeologic study, as well as how to carry out and report on such work. However, it seems appropriate to end the book with a few final suggestions. Some reiterate and generalize the main points made previously. Others treat topics not covered by any of the chapters, but important in hydrogeologic work today.

REMEMBER THE BASICS

One of the main objectives of this book has been to remind everyone to start at the beginning and carefully proceed through all the required steps when conducting hydrogeologic studies, regardless of the purpose. This process involves not only compiling existing data from published sources and files of government agencies or making new field observations, but then synthesizing such information into a sound conceptual hydrogeologic model of the study area. You should not start drilling a supply well, chasing a contaminant plume, or making a numerical model before you have done so. In other words, always do your homework.

Also, provide as much fundamental information in reports as possible and as needed. If the study cries out for a certain kind of illustration, and the data permit it, provide one. However, it is not enough to provide an illustration; it must be correct and complete. For example, meaningful decisions cannot be made from a water-table map that mixes unconfined, confined, and composite water levels. Obviously, a map with wells plotted inaccurately, no location grid, no scale, etc., will also be of little use.

Become familiar with the terminology of the industry and use it correctly. This applies to geologic and hydrologic terms alike. For example, be careful to employ formal stratigraphic nomenclature properly. Also, know when to use *aquifer* and when to use *saturated zone*. Don't say *gradient* when you mean *flow direction*.

If you review hydrogeologic reports prepared by others, and necessary materials are not included, require that they be added. This will not only improve the product at hand, but also the skills of the author and future reports. If the required material is not added, you will deal with their inadequate reports indefinitely. By approaching this as a clarification of your report requirements, rather than a criticism of the author, feelings will be spared.

An important part of solving many of today's water problems is public education. If your employer holds or participates in public meetings or hearings, volunteer to make a presentation on hydrology in general or on your study area. Since you cannot explain something you do not understand, this will give you an excellent opportunity to practice using basic hydrogeologic concepts and terminology correctly. At the very least, you will gain some experience in public speaking while improving public perception of hydrology or a local environmental topic.

USE INTUITION

When you characterize hydrogeologic systems, all the information you need may not be readily available. In such cases, some assumptions must be made. These need not be wild speculations, but rather reasonable suppositions, based on what is known and what can logically be expected.

For example, consider the task of contouring water levels. The first thing to determine is the general shape or trend of the contours. Use topography and site history, together with available data, to decide this. Water-level contours mimic the terrain. Furthermore, if the area contains a source of artificial recharge (lagoon, outfall, infiltration field, etc.), there is likely to be a ground-water mound, illustrated by contours bulging down-gradient. By contrast, if the site includes production wells, there may be a cone of depression or ground-water trough, shown by contours bulging up-gradient. Because paleotopography along a disconformity can also influence the configuration of the water table and flow direction, take that into account in contouring.

APPLY MODELS WISELY

Numerical models are powerful hydrogeologic tools when used properly. If you should do some modeling, avoid the common pitfalls of overkill, misconceptualization, inappropriate assignment of parameters, and blind faith. Also watch for these in any model you are called on to review or use. Be slow to accept a dubious model solution to a problem that could easily be solved by a well in a key location or some other means of collecting actual data. Remember, models are intended to complement skills, not replace them.

BE RIGHT

Many hydrologic studies are conducted in response to regulatory action on some issue (e.g., water rights, a proposed waste-disposal site, ground-water contamination, etc). Thus, regardless of your position, it is important to have all the facts and figures and that they be correct. Search out all previous works that apply. Compile the best data set possible. Formulate a sound conceptual hydrogeologic model for the area. Where additional data are needed, try to obtain them.

It is not enough to have some data, they must be correct. Assume that the other side will look for mistakes in your work and use them to discredit your position. Thus, it is important to check your work carefully. This may involve making calculations twice, having a coworker read from a table while you check values on a map, or getting peer reviews of your reports. If you have any doubts about your interpretations, check some standard references or discuss them with a senior coworker.

Documentation of competence is often required. There has long been a requirement that engineers be registered. Now, more and more states are requiring that hydrogeologists be registered or certified as well. Even if documentation is not required, having it makes you more competitive, especially for consulting jobs. Some companies require it. Thus, when you are confident that your training and experience warrant it, you should apply for professional registration or certification. Some states have their own programs; others recognize certification by various organizations, such as the National Ground Water Association (on-line at http://www.ngwa.org) and the American Institute of Hydrology (http://www.aihydro.org).

MAINTAIN PROFESSIONALISM

In cases that involve regulatory action, there is a natural tendency on the part of the responsible party to be defensive, and in some cases, lash out at those trying to remedy the problem. The working climate can easily become adversarial. Thus, on occasion, regardless of your position, you may find yourself in a potentially confrontational meeting. This is unfortunate and obviously not conducive to resolving the problem. However, you must do your best, despite such conditions.

If you are representing the applicant in a water-rights case or the responsible party in a ground-water–contamination case, avoid the temptation to ambush the regulators. If you have done your work carefully, you can present and defend it confidently. Try to remember that if the regulators find it lacking, it is not due to some flaw in their character or breeding. Rather, the work you have presented may not be based on good science or some key facts are unclear to them or are missing.

If you are representing the regulator, be thoroughly prepared. That is, identify the real issues, have the pertinent facts at hand, and anticipate the stand likely to be taken by the other side. The best defense against an ambush is to have a specific sequence of critical questions ready and at the first opportunity lead the discussion in that direction, rather than let it get bogged down in a defensive duel of egos.

At all costs, avoid the temptation to respond in kind to petty attacks. In short, to avoid an unpleasant (and unproductive) situation, maintain your professionalism at all times.

PRACTICE GOOD WRITING

The best way to improve any skill is to practice it. This is as true for technical writing as for any other endeavor. A useful approach is to apply the rules of good style and strive for clarity, no matter what you are writing. As you read journal articles or review documents, watch for examples of good scientific or technical prose, and imitate them in your own writing. In committing your thoughts to paper, consider how they would be stated in one of the journals you have read. With every memo and letter you compose, your writing will improve. Eventually you will find even the major reports are not the painful chores they once seemed.

In reporting your findings, consider how you would like them presented if you were the user. Don't assume readers will go through your report cover-to-cover. Can major topics of interest be readily located? Make sure that they can be by carefully planning the organization of the text and reflecting this in a useful table of contents.

KEEP LEARNING

Hydrogeology is a complex and ever-changing field. You must constantly upgrade your skills through various educational opportunities. These include formal college courses, technical shortcourses, membership in professional organizations, participation in professional conferences, attendance at technical lectures, as well as reading technical reports and manuals.

The number of new water-management and environmental graduate programs increases yearly. Many employers provide financial support and release time for their staff members to pursue advanced training or even degrees. If there is a university in your area, check into available programs. If you are not near a university, the Universities Council on Water Resources (Urban and Hubbard 1994) has prepared a booklet listing schools that provide training in hydrology and the courses they offer. The list includes graduate programs in water resources available in the United States. The booklet is in its ninth edition and is periodically updated. It is also available on-line at http://www.uwin.siu.edu/ucowr/grad/.

Even if you are not pursuing a degree, you may benefit from taking some college courses in or related to the hydrological sciences. Many such courses are offered in the evening so that working professionals can take advantage of them. However, if you cannot give such courses the attention they require, it is better to defer enrollment until you can.

If formal college courses are not available in your area or are difficult to work into your schedule, consider technical short courses. Various professional organizations, agencies, and even consultants offer these on a various classic or timely topics, such as aquifer analysis (e.g., Senay 1987) or vadose-zone monitoring (e.g., Everett and Kreamer 1990). Environmental Education Enterprises (FAX 614 792-0006) and the National Ground Water Association, both based in Columbus, Ohio, offer short courses on a wide range of hydrologic and environmental topics. If a number of people would benefit, but travel funds are limited, it may be possible to bring the training to your place of employment. Similarly, if special training is needed, but the topic is not among those listed in the offerings of the provider, trainers may be willing to develop a custom course, especially if there is a group to take it. Professional short courses vary in length, cost, and focus. However, there is bound to be one you can afford on a topic you can use. When pertinent to your work, your employer may even pay for such courses.

Joining one of the professional societies devoted to hydroscience is another good way to keep up to date. Most have membership information on-line (see addresses previously listed for the National Ground Water Association and American Institute of Hydrology). Some even have state chapters or sections (e.g., the American Water Resources Association). These address local or regional issues and put you in touch with other hydrologists in your area. The technical journals that often accompany membership highlight new techniques and give case histories, some perhaps pertinent to your own projects. These professional articles are good examples of technical writing that you can consult when preparing your own reports or works for publication. Professional organizations may also be able to help you locate the kind of hydroscience program or short courses you are interested in. They also list conferences and employment opportunities.

Conferences, regularly held by professional societies, universities, and government agencies are an excellent way to stay on top of various aspects in your field. These may focus on local geologic conditions and water issues, for example, the Albuquerque Basin, New Mexico (New Mexico Water Resources Research Institute 1995) or be of international scope and participation, for example, low-permeability rocks (International Association of Hydrologeologists 1985). The calendar section of *Geotimes,* the monthly magazine of the American Geological Institute, regularly lists conferences and workshops of interest to a wide range of geoscientists. This includes announcements of both international and national events. Although a given issue only lists events for the current and the following month, a complete events calendar and a new events section is available on the World Wide Web at http://www.agiweb.org/agi/geotimes.html. Publications of other organizations may at least list national conferences.

Lectures at universities serve a function similar to professional conferences. If you are near a campus, check with the geoscience or civil engineering departments to see if a series of pertinent talks is offered, and pick up a schedule. Because universities are also the venue for various traveling speakers sponsored by the professional organizations, you may want to get on the mailing or e-mail list for announcements of such special lectures.

Technical manuals prepared and distributed by government agencies are also very instructive when you are building your hydrogeologic skills. For example, some of those by the USGS and EPA (listed in chapters 11 and 12) have become standard references for the working hydrologist. Some of the water-supply papers of the USGS, especially the earlier ones, are devoted to specific hydrologic field methods or analytical techniques, for example, those on aquifer tests by Ferris and others (1962) and well logging by Patten and Bennett (1963).

The Internet is the latest way to keep up with your field. Home pages for government agencies announce new publications, public meetings, etc., that may be pertinent to your work. Home pages for the professional organizations not only give membership information, but also calendars of conferences and short courses. Entries for universities give information on programs of study, faculty specialties, and current research topics. Some useful hydrogeology addresses are maintained by interested professionals.

In short, as you conduct your hydrogeologic studies, try to remember the basics, use your intuition, apply models wisely, be right, maintain professionalism at all times, practice good writing, and, above all, keep learning!

REFERENCES

Everett, L. G., and D. K. Kreamer. 1990. Vadose zone monitoring techniques. Course notes.. National Water Well Association.

Ferris, J. G., D. B. Knowles, R. H. Brown, and R. W. Stallman. 1962. Theory of aquifer tests. Water-supply paper 1536-E. U.S. Geological Survey. 174 p.

International Association of Hydrogeologists. 1985. Hydrology of rocks of low permeability. *Memoires XVII.* Proceedings of 17th International Congress, International Association of Hydrogeologists. 862 p.

New Mexico Water Resources Research Institute. 1995. The water future of Albuquerque and the middle Rio Grande basin. Report no 290. Proceedings of 39th Annual New Mexico Water Conference. New Mexico Water Resources Research Institute. 449 p.

Patten, E. P., Jr, and G. D. Bennett. 1963. Application of electrical and radioactive well logging to ground-water hydrology. Water-supply paper 1544-D. U.S. Geological Survey. 60 p.

Senay, Y. 1987. Aquifer analysis. Course notes. National Water Well Association.

Urban, L. V., and J. E. Hubbard. 1994. Graduate studies in water resources. The Universities Council on Water Resources, Carbondale, IL. 56 p.

Glossary

Alluvial deposited by running water on broad slopes or aprons (fans), or in basins adjacent to ranges.

Alluvium alluvial deposit; usually unconsolidated mixture of gravel, sand, silt, and clay.

Angular unconformity an unconformity across which strata are not parallel.

Aquifer geologic unit whose saturated portion yields significant quantities of water to wells or springs; material which both stores and transmits water.

Aquitard (also Confining bed) geologic material of relatively low hydraulic conductivity overlying an artesian aquifer and responsible for the confinement of water within it; material which stores, but only poorly transmits water.

Artesian (also Confined) term applied to ground water that rises above the level at which it is encountered during drilling; also applied to wells in which this rise occurs and aquifers that produce it. The rise is not necessarily to the ground surface; in that case the term "flowing artesian well" is applied.

Calcite mineral composed of calcium carbonate ($CaCO_3$); main mineral in limestone.

Carbonaceous said of a rock containing carbon or carbonized organic (mainly vegetal) material.

Clastic term applied to sedimentary deposits or rocks composed of fragments of rock or mineral matter of various sizes.

Coal measures succession of sedimentary rocks associated with coal: sandstone, siltstone, and shale or claystone.

Colluvium deposit formed on or at the base of slopes by gravity or unconcentrated runoff; usually unconsolidated rock fragments and soil.

Contact boundary between adjacent stratigraphic units.

Dewatering removal of water by pumping to lower the water table at a mine or construction site.

Discharge movement of water out of an aquifer; process by which ground water is depleted; may be natural or artificial (pumping). Also the disposal of effluent or liquid waste.

Disconformity unconformity along which there is erosional relief.

Drawdown lowering of the water table or potentiometric surface for an aquifer in response to pumpage or flow from artesian wells.

Elevation head the elevation of a point in the saturated zone above a datum.

Eolian deposited by the action of the wind.

Ephemeral said of a stream that flows only in direct response to precipitation or storm runoff in the vicinity and whose channel is usually above water table; also the flow of such a stream.

Evapotranspiration combined loss of subsurface water to the atmosphere through the processes of evaporation from soil and transpiration by plants.

Fault fracture or break in geologic media along which there has been movement.

Fluvial deposited by running water in discrete channels, as with rivers and streams.

Formation fundamental unit in the local stratigraphic classification of rocks, used on geologic maps and in cross sections; larger than member and smaller than group. Must be mappable at the scale of common topographic quadrangles.

Ground water subsurface water in the zone of saturation; that water below the water table.

Group stratigraphic unit made up of two or more formations.

Gypsum mineral composed of calcium sulfate ($CaSO_4$); may occur in layers with limestone, shale, or other evaporites.

Head the elevation of the water level in a well above a specified datum; more specifically, see Elevation, Hydraulic, Static, Total, Velocity head

Hydraulic conductivity volume of water (at existing viscosity) that will move in unit time, under a unit hydraulic gradient, through a unit area of saturated material; traditionally reported as gpd/ft^2. If gal are converted to ft^3, unit becomes ft/d, as a result of algebraic cancellation.

Hydraulic gradient change in static head per unit of distance in a given direction; units are given if different (ft/mi) or omitted if the same.

Hydraulic head (also Static head) height (above a datum) of a column of water that can be supported by the static fluid pressure at a given point; the sum of the elevation head and pressure head.

Igneous rock one formed by cooling of molten rock material; may be produced at the surface or in the subsurface.

Intermittent said of a stream along which perennial flow is restricted to certain reaches; also the flow of such a stream.

Limestone sedimentary rock consisting of >50% calcite.

Listric fault one with a concave-upward surface that curves up from horizontal gently at first, then more steeply.

Lithology physical character of a rock or deposit expressed in terms of texture, mineralogy, color, and thickness.

Member stratigraphic unit representing a subdivision of a formation.

Metamorphic rock one formed by metamorphism, that is, the alteration of a preexisting rock through increased temperature, pressure, and changes in chemistry.

Mineral naturally occurring, inorganic substance, with a characteristic set of physical properties and a fixed chemical composition or range of composition; the basic components of rocks.

Nonconformity unconformity between rocks of different major types (for example, igneous and sedimentary).

Nonclastic term applied to crystalline sedimentary rocks composed of chemically or biologically precipitated minerals.

Perennial said of a stream that flows year round; also the flow of such a stream.

Permeability (also Intrinsic permeability, specific permeability) measure of the relative ease with which a porous medium transmits a liquid; a property of the medium alone, independent of liquid properties or forces causing movement.

Piezometric surface surface that represents static head from water-level measurements in piezometers installed in either unconfined or confined aquifers.

Porosity portion (percent) of the total volume of rock, unconsolidated sediment, or soil taken up by open space or pores; equal to the sum of specific yield and specific retention.

Potentiometric surface level to which ground water will rise in wells; surface that represents the static head for a given aquifer.

Pressure head hydrostatic pressure expressed as the height of a column of water the pressure can support.

Pumping test (also Aquifer test or well test) test of a well to determine the yield or hydraulic properties of the aquifer penetrated; involves regularly observing water-level drawdown during pumping at a known rate; may also involve observing recovery after pumping or injecting a known volume of water (or a solid).

Quartz mineral composed of crystalline silica or silicon dioxide (SiO_2).

Recharge movement of water into an aquifer; process by which aquifers are replenished; may be natural or artificial (injection, flooding).

Rock naturally occurring aggregate of minerals.

Sandstone clastic sedimentary rock composed mainly of sand-sized particles of mineral or rock material.

Sedimentary rock one formed by the deposition of physical (clastic) or chemical (nonclastic) sediments.

Shale clastic sedimentary rock composed mainly of clay, having a tendency to split into thin, platy layers.

Soil surficial material characterized by various distinct horizons; the result of soil-forming processes acting on unconsolidated sediment or weathered rock.

Soil water subsurface water in soil or rock within the unsaturated zone; that above the water table.

Specific capacity the discharge of a well divided by the drawdown in it; measured as gpd/ft of drawdown. If gal are converted to ft^3, unit becomes ft^2/d through algebraic cancellation.

Specific conductance electrical measure of a water's dissolved solids or salt content; the reciprocal of resistance. Measured as microSiemens per centimeter (μS/cm).

Specific retention volume of water per unit volume of porous medium that the porous medium will retain after drainage by gravity flow; equal to porosity minus specific yield.

Specific yield volume of water per unit volume of porous medium that will drain from the medium under the influence of gravity; equal to porosity minus specific retention.

Storativity volume of water released from or taken into storage per unit surface area of aquifer per unit change in hydraulic head; dimensionless parameter.

Specific storage (also Elastic storage coefficient) amount of water per unit volume of a saturated medium that is stored or released due to compressibility of the medium and pore water per unit change in head.

Total head the sum of elevation head, pressure head, and velocity head.

Total dissolved solids physical measure of salinity; amount (mg/L) of residue obtained by oven drying a water sample. Water quality may be generally classified by this parameter:

TDS (mg/L)	Classification
<1,000	fresh
1,000-3,000	slightly saline
3,000-10,000	saline
10,000-35,000	very saline
>35,000	brine.

Transmissivity rate at which water (at existing viscosity) is transmitted through a cross section of material having the dimensions unit width and total thickness as height, under a unit hydraulic gradient; hydraulic conductivity times the thickness of the aquifer. It should be noted that transmissivity is not defined in the vertical dimension. Measured as gpd/ft of width; if gal converted to ft^3, unit becomes ft^2/d through algebraic cancellation.

Unconformable said of a contact between stratigraphic units across which there is a gap in the rock record; that is, some interval of geologic time is not represented.

Unconformity unconformable contact (see also Angular unconformity, disconformity, nonconformity).

Velocity head energy of flow expressed as vertical distance through which a fluid would fall in order to attain the given velocity; equal to $v^2/2g$, where v is ground-water velocity and g is the acceleration due to gravity.

Viscosity property of a fluid that determines its ability to resist flow; dependent on temperature and density.

Water level used to refer to both the depth and the elevation of water standing in a well; applied to unconfined and confined water alike.

Water table surface that represents the static head for a given unconfined aquifer; the top of the saturated zone; the surface formed by points at which water pressure equals atmospheric pressure.

APPENDIX A

Guide to Logging Cuttings/Core

Samples of cuttings and core are an important source of geologic and hydrologic information. Logs should at least be made in the field during drilling and later in the laboratory/office, if possible. In both cases, the logs should be prepared in a consistent style. The purpose of this guide is to set forth the basic parameters to be described in a log, as well as a suggested log format. An example of the type of log you should strive for is also given for reference.

HEADING

Include several basic items at the top of the log:

Well/borehole name or number

Location: quarters, sec, twp, rge; topographic quadrangle; county; distance/direction from familiar landmark; property owner

Elevation: ___ ft (how determined)

Date drilled:

Method/equipment:

Company/driller:

Completed:

Construction/abandonment:

Site geologist/logger:

Others logging:

DESCRIPTION

Each material or unit should be described using a standard list of parameters in a set order:

Main constituent (general name: gravel, silty clay, etc.)

Color (name and number designation from standard color chart)

Grain size (texture of main constituent: fine; medium-coarse, etc.)

Composition (main rocks/minerals present)

Minor constituents (like for main constituent above)

Other (reaction to HCl, cement type, etc., especially if core)

Remarks (drilling notes; special analyses/tests)

Note first water and water-bearing intervals as they are encountered and indicate total depth drilled at end. If it is a hollow-stem-auger core hole, identify no-recovery intervals.

Example

Borehole No. MW1E

Location: NW, NE, SW, 14, 11N, 3E; Deseret Quadrangle; Bernardo County; approximately 1 mile east of intersection of Sweet Water Blvd. and I-60, east of tank farm; Black Oil Refinery

Elevation: 4,983 Ft (surveyed)

Drilled: 25 Dec 1994–1 Jan 1995

Method/Equipment: Air rotary; truck-mounted Wal-Mart 007

Company/driller: Corkscrew Drilling; Lefty LaToole

Completed: 14 Feb 1995

Construction/abandonment: completed as 2″ monitoring well (screen at 95-110 ft; cement from 0-5 ft, volclay from 5-90 and 115-120 ft, filter pack from 90-115 ft).

Site geologist: I. C. Awl, State EPA

Others logging: Bye D. Owher, Allso-Rann Consulting

Depth (ft)	Lithologic Description
0–3.5	FILL—brown (7.5YR4-5/2) to dark grayish brown (10YR4/2), pebbly sandy loam
3.5–5	Not sampled
5–25	CLAY—olive (5Y4/3); Driller-smooth drilling
25–37	SILTY CLAY—as above but silty
37–52	SAND—pale brown (10YR6/3), very fine, uniform, mainly quartz
52–95	PEBBLY SAND—reddish brown (5YR5/3), sand medium-coarse, mostly quartz and feldspar, some fine-grained rock fragments; whole pebbles up to 1/4 in, mixed volcanic compositions; trace of carbonate coatings on some pebbles
95–120	aa, but pebbles more abundant and larger (chips larger than 1/4 in) and sand up to very coarse; first water at 100 ft (remained this level), abundant
TD = 120	

APPENDIX B

Well-Inventory Data Sheet

Date:________ 19____, **Basin:**____________ **File:** _______________, Well No:______

Recorded By: ________________________ **Source:** (circle one) Field Observation / interview

1. **Location:** Grid: ____________________, Northing:_______, Easting: _______
   ____ 1/4, ____ 1/4, ____ 1/4, Sec. ____, T._____, R._____, County: __________
   ____ Miles ___ (direction) of ______________ (landmark): Topo. Quad. ________
   Access:____

2. **Owner/Tenant:** ________________, Address: ________________________

3. **Setting** Topography: ____________________________, Elevation: ________ft.
   Surface Geology: ____

4. **Use:** Dom. / Stock Public/Sup. / Ind. / Irrig. / Mon. / Piezo. / Not in Use / Abandoned

5. **Drilled by:** ________________ Address: __________________ Date:_______
   Total Depth Drilled: Reported/Measured: ______ ft. Plugged back to ______ ft.

6. **Initial Water Level:** Reported/Measured ______ ft. Above/Below ______ Date: _______
   Static Water Level: Reported/Measured ______ ft. Above/Below ______ Date:_______
   Aquifer: ________________

7. **Current Water Level:** Reported/Measured: ___ ft. Above/Below Measuring Pt.
   which is ______________________ at ____ ft. Above Below Ground Surface.

8. **Construction:** Casing:___ in., Material __________, Riser: ___ in., Material ________
   Screen: Interval(s):____________________________________, Type ________
   Filter Pack:____________________________ From _______ ft. To _______ ft.

9. **Pump:** Type:________________ Capacity: ________ gpm, Power ____________
   Make and Model _________________, Horsepower:_____, Drop Pipe Diameter:_____, in.

10. **Yield:** Flowing/Pumping _____ gpm Estimated/Reported
    Measuring Method: ____

11. **Quality:** Specific Conductance___________ microsiemens/cm, Temp.: °F/°C: ________
    pH: ______ Odor: ______________________ Color: ______________
    Sample (✓)?____, No.:______ Other: ______________________________

12. **Remarks:** ____
    ____
    ____

APPENDIX C

Elements of an Ideal Conceptual Hydrogeologic Model

Geology

- Stratigraphy
  - Information
    - the sequence
    - units—their lithology, thickness, extent, variability
  - Illustrations
    - geologic column
    - measured section(s)
    - well log(s)
    - cross section(s)
    - depth map(s)—to top of unit(s)
    - thickness map(s)—of unit(s) or interval(s) of interest
- Structure
  - Information
    - dip/strike
    - folds
    - faults
    - fracture/joint systems
  - Illustrations
    - geologic map—with dip, strike, folds, faults
    - cross section(s)
    - structure map(s)—elevation of top of unit
- Geomorphology
  - Information
    - relief
    - major landforms
  - Illustrations
    - topographic map
    - aerial photograph(s)
    - ground-level photograph(s)

Hydrology

- Surface Water
  - Information

- major features (rivers, lakes, etc.)
- properties
- relationship to ground water

Illustrations

- hydrograph(s)
- flow-frequency plot(s)
- gain/loss plot(s)
- rating curve—stage versus discharge
- flow data table
- longitudinal profile
- channel cross section(s)
- water-quality table(s)

Soil Water

Information

- thickness of vadose zone
- nature of the medium
- moisture content
- solute chemistry

Illustrations

- profiles (moisture content/chemistry/texture versus depth)
- maps (thickness of vadose zone, average moisture content, moisture at a certain depth, average or maximum concentration of a given solute)
- logs (records of any boreholes in vadose zone)

Ground Water

Information

- aquifer(s)
- their characteristics
- water depth
- regional or perched?
- unconfined or confined?
- recharge (area, mechanism, rate)
- flow (direction, rate, gradient)
- discharge (area, mechanism, rate)
- water quality (constituents, variations across area)

Illustrations

- well-records table (water-well data)
- well-construction diagrams
- water-level map(s)—contours/flow directions for various aquifers, if different
- water-depth map—depth to regional water table, or first water
- water-quality table(s)—hydrochemical data
- water quality maps (water type, concentration contours)
- water-quality diagrams (Piper, Stiff, etc)

Synthesis

Information

- Any major geologic differences across area?
- Relationship of surface, soil, and ground waters?
- Water budget for area?
- Geologic controls of hydrologic phenomena?

Illustrations

- hydrogeologic column—relationship of stratigraphic and hydrostratigraphic units, their water-yielding properties
- hydrogeologic cross section—stratigraphic units, hydrologic function (aquifer/aquitard), water level, equipotential contours, flow direction(s)

APPENDIX D

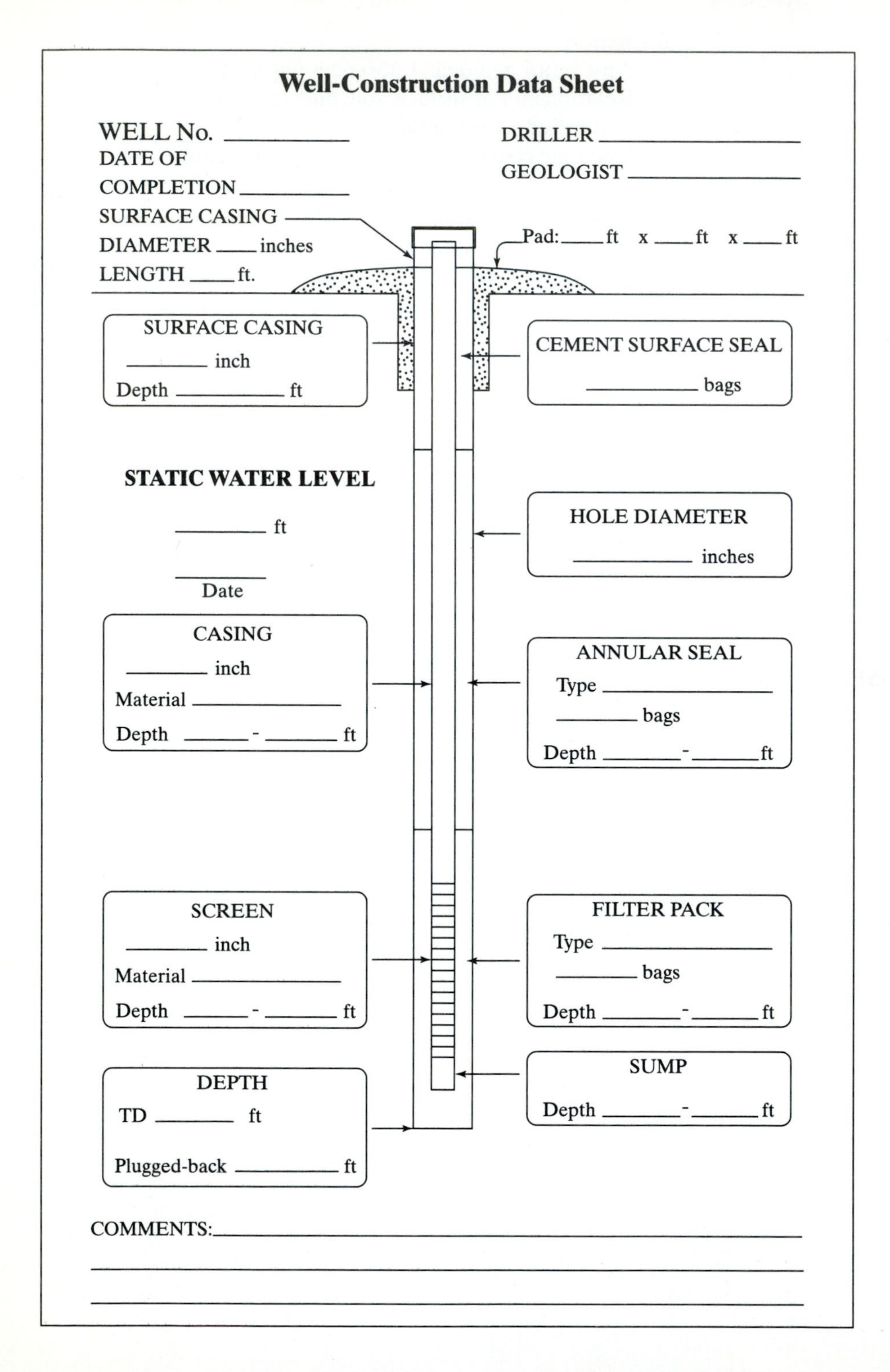

Well-Construction Data Sheet
WELL No. ____
DRILLER ____
DATE OF COMPLETION ____
GEOLOGIST ____
SURFACE CASING
DIAMETER ___ inches
LENGTH ___ ft.
Pad: ___ ft x ___ ft x ___ ft
SURFACE CASING
___ inch
Depth ___ ft
CEMENT SURFACE SEAL
___ bags
STATIC WATER LEVEL
___ ft
Date
HOLE DIAMETER
___ inches
CASING
___ inch
Material ___
Depth ___ - ___ ft
ANNULAR SEAL
Type ___
___ bags
Depth ___ - ___ ft
SCREEN
___ inch
Material ___
Depth ___ - ___ ft
FILTER PACK
Type ___
___ bags
Depth ___ - ___ ft
DEPTH
TD ___ ft
Plugged-back ___ ft
SUMP
Depth ___ - ___ ft
COMMENTS: ___

APPENDIX E

Miscellaneous Conversions, Equivalents, and Formulas

Useful Conversions within FPS System

	To Convert	Into	Multiply By
Length	mile	ft	5,280
Area	acre	ft^2	43,560
Volume	ft^3	gal (US)	7.48
	gal (US)	ft^3	0.134
	ac-ft	ft^3	43,560
	ac-ft	gal (US)	325,900
Discharge	cfs	gpm	448.831
	cfs	ac-ft/d	1.98
Time	day	min	1,440
	day	sec	86,400
	year	min	525,600
	year	sec	31,536,000

USEFUL EQUIVALENTS

1 in. of rain = 27,200 gal/ac

1 ft of water = 0.4335 psi

1 psi = 2.307 ft of water

1 ft of water = 0.8826 inch of mercury

1 in. of mercury = 1.1333 ft of water

1 ft of water = 0.0295 atmospheres

1 ft^3 of water = 62.5 lbs

1 ft^3 of water = 7.48 gal

1 gal of water = 8.336 lbs

1 miners in. = 12 gpm (may differ from state to state)

USEFUL FORMULAS

Annulus Volume (for filter pack, seal or fill)

$V_a = (\pi r_2^2 h) - (\pi r_1^2 h)$, where r_2 = radius of larger casing (inside), r_1 = radius of smaller casing (to outside) and h = casing length

Casing Volume (for purging)

$V_c = \pi r^2 h$, where r = radius of casing (inside) and h = casing length

Channel cross-sectional area (trapezoid)

$A = (a + b) / 2 \times h$, where a and b are top and botton channel widths and h is channel depth

Index

H

I

J

K

L

M

N

O

P

Q

R

S

T

U

V

W

X